STERLING
Test Prep

College Biology

Cell & Molecular Biology

Questions & Explanations

4th edition

Customer Satisfaction Guarantee

Your feedback is important because we strive to provide the highest quality prep materials. Email us questions, comments or suggestions.

info@sterling–prep.com

We reply to emails – check your spam folder

4 3 2 1

ISBN-13: 979-8-8855714-5-6

Sterling Test Prep materials are available at quantity discounts.

Contact info@sterling–prep.com

Sterling Test Prep
6 Liberty Square #11
Boston, MA 02109

Published by Sterling Test Prep

Printed in the U.S.A.

STERLING
Test Prep

Thousands of students use our study aids to achieve higher grades!

To achieve a high grade in college biology, you need to do well on your tests and final exam. This book helps you develop and apply knowledge to quickly choose the correct answers to cell and molecular biology questions typically tested in college courses. Solving targeted practice questions builds your understanding of fundamental concepts and is a more effective strategy than merely memorizing terms.

This book has 655 practice questions covering cell and molecular biology topics. Biology instructors with years of teaching experience prepared this practice material to build your knowledge and skills crucial for success in a college biology course. Our editorial team reviewed and systematized the content for targeted preparation.

The detailed explanations describe why an answer is correct and – more important for your learning – why another attractive choice is wrong. They teach the scientific foundations and details of essential cell and molecular biology topics needed to answer questions of varying level of difficulty. Read the explanations carefully to understand how they apply to the question and learn important biology principles and the relationships between them. With the practice material contained in this book, you will significantly improve your understanding, test scores, and your grade.

We wish you great success in your academic achievements and look forward to being an important part of your preparation!

250926akp

College Cell & Molecular Biology Subject Review provides comprehensive coverage of cell and molecular biology topics taught in standard college course. The content covers all essential principles and concepts necessary to understand the material and do well on college tests and exams.

Visit our Amazon store

College Organismal Biology:
Evolution, Ecology, Structure and Function Systems

If you benefited from this book, please leave a review on Amazon so others can learn from your input. Reviews help us understand our customers' needs and experiences while keeping our commitment to quality.

Table of Contents

College study aids by Sterling Test Prep

Cell & Molecular Biology
Organismal Biology
Physics First Semester
Physics Second Semester
General Chemistry
Organic Chemistry I
Organic Chemistry II
Physical Chemistry
Anatomy & Physiology

Environmental Science
Human Geography
Psychology
Sociology
United States History
American Government & Politics
Comparative Government & Politics
World History
European History

Visit our Amazon store

Cell & Molecular Biology

Practice Questions

Page intentionally left blank

Section I: Cell Structure & Function

Eukaryotic Cell

1. Facilitated transport can be differentiated from active transport because:

A. active transport requires a symport
B. facilitated transport displays saturation kinetics
C. active transport displays sigmoidal kinetics
D. active transport requires an energy source
E. active transport only occurs in the mitochondrial inner membrane

2. The cell is the basic unit of function and reproduction because:

A. subcellular components cannot regenerate whole cells
B. cells can move in space
C. single cells can sometimes produce an entire organism
D. cells can transform energy to do work
E. new cells can arise by the fusion of two cells

3. Which is/are NOT able to readily diffuse through plasma membranes without a transport protein?

I. water
II. small hydrophobic molecules
III. small ions
IV. neutral gas molecules

A. I only
B. I and II only
C. I and IV only
D. III and IV only
E. I and III only

4. During a hydropathy analysis of a recently sequenced protein, a researcher discovers that the protein has several regions containing 20-25 hydrophobic amino acids. What conclusion would she draw from this finding?

A. Protein would be specifically localized in the mitochondrial inner membrane
B. Protein would be targeted to the mitochondrion
C. Protein is likely an integral protein
D. Protein is likely involved in glycolysis
E. Protein is likely secreted from the cell

5. If a membrane-bound vesicle that contains hydrolytic enzymes is isolated, it is likely a:

A. vacuole
B. microbody
C. chloroplast
D. phagosome
E. lysosome

6. A DNA damage checkpoint arrests cells in:

A. M/G2 transition
B. G1/G2 transition
C. G1/S transition
D. anaphase
E. S/G1 transition

7. If a segment of double-stranded DNA has a low ratio of guanine-cytosine (G-C) pairs relative to adenine-thymine (A-T) pairs, it is reasonable to assume that this nucleotide segment:

A. requires less energy to separate the two DNA strands than a comparable segment with a high C-G ratio
B. requires more energy to separate the two DNA strands than a comparable segment with a high C-G ratio
C. requires the same energy to separate the two DNA strands as a comparable segment with a high C-G ratio
D. contains more adenine than thymine
E. contains more cytosine than guanine

8. In general, phospholipids contain:

A. a glycerol molecule
B. saturated fatty acids
C. unsaturated fatty acids
D. a cholesterol molecule
E. a glucose molecule

9. The overall shape of a cell is determined by its:

A. cell membrane
B. cytoskeleton
C. nucleus
D. cytosol
E. endoplasmic reticulum

10. Which molecule generates the greatest osmotic pressure when placed into water?

A. 300 m*M* NaCl
B. 250 m*M* $CaCl_2$
C. 500 m*M* glucose
D. 600 m*M* urea
E. 100 m*M* KCl

11. Which cellular substituent is produced within the nucleus?

A. Golgi apparatus
B. lysosome
C. ribosome
D. rough endoplasmic reticulum
E. cell membrane

12. In the initial stages of the cell cycle, progression from one phase to the next is controlled by:

A. p53 transcription factors
B. anaphase-promoting complexes
C. origin recognition complexes
D. pre-replication complexes
E. cyclin–CDK complexes

13. Which is the correct sequence occurring during polypeptide synthesis?

A. DNA generates tRNA → tRNA anticodon binds to the mRNA codon in the cytoplasm → tRNA is carried by mRNA to the ribosomes, causing amino acids to join in a specific order
B. DNA generates mRNA → mRNA moves to the ribosome → tRNA anticodon binds to the mRNA codon, causing amino acids to join in their appropriate order
C. Specific RNA codons cause amino acids to line up in a specific order → tRNA anticodon attaches to mRNA codon → rRNA codon causes the protein to cleave into specific amino acids
D. DNA regenerates mRNA in the nucleus → mRNA moves to the cytoplasm and attaches to the tRNA anticodon → operon regulates the sequence of amino acids in the appropriate order
E. DNA generates mRNA → mRNA anticodon binds to tRNA codon causing amino acids to join together in their appropriate order

14. Both prokaryotes and eukaryotes contain:

I. a plasma membrane   II. ribosomes   III. peroxisomes

A. I only
B. II only
C. I and II only
D. II and III only
E. I, II and III

15. RNA is NOT expected to be in a:

A. nucleus
B. mitochondrion
C. prokaryotic cell
D. ribosome
E. vacuole

16. Phosphotransferase is needed to form the mannose-6-phosphate tag that targets hydrolase enzymes to their lysosomal destination. Defective phosphotransferase causes I-cell disease, whereby the defective organelle which gives rise to this condition is the:

A. nucleus
B. cell membrane
C. Golgi apparatus
D. smooth ER
E. nucleolus

17. Mitochondria and chloroplasts are unusual organelles because they:

A. synthesize all their ATP using substrate-level phosphorylation
B. contain cytochrome C oxidase
C. are devoid of heme-containing proteins
D. contain nuclear-encoded and organelle-encoded proteins
E. degrade macromolecules using hydrolytic enzymes

18. All statements are true about cytoskeleton EXCEPT that it:

A. is not required for mitosis
B. maintains the cell's shape
C. gives the cell mechanical support
D. is composed of microtubules and microfilaments
E. is important for cell motility

19. Inside the cell, several key events in the cell cycle include:

I. DNA damage repair and replication completion
II. centrosome duplication
III. assembly of the spindle and attachment of the kinetochores to the spindle

A. I only
B. I, II and III
C. I and II only
D. II and III only
E. I and III only

20. A researcher labeled *Neurospora* mitochondria with a radioactive phosphatidylcholine membrane component and followed cell division by autoradiography in an unlabeled medium, allowing enough time for one cell division. What results did this scientist observe before concluding that pre-existing mitochondria give rise to new mitochondria?

A. Daughter mitochondria are labeled equally
B. Daughter mitochondria are all unlabeled
C. One-fourth of daughter mitochondria are labeled
D. Some of the daughter mitochondria are unlabeled, while some are labeled
E. Daughter mitochondria are all labeled

21. Which involves the post-translational modification of proteins?

A. peroxisomes
B. vacuoles
C. Golgi complex
D. lysosomes
E. smooth ER

22. The width of a typical animal cell is closest to:

A. 1 millimeter
B. 20 micrometers
C. 1 micrometer
D. 10 nanometers
E. 100 micrometers

23. Which organelle is identified by the sedimentation coefficient – S units (Svedberg units)?

A. peroxisome
B. nucleus
C. mitochondrion
D. nucleolus
E. ribosome

24. Which is NOT involved in osmosis?

A. H_2O spontaneously moves from a hypertonic to a hypotonic environment
B. H_2O spontaneously moves from an area of high solvent to a low solvent concentration
C. H_2O spontaneously moves from a hypotonic to a hypertonic environment
D. Transport of H_2O
E. Diffusion of H_2O

25. During cell division, cyclin B is marked for destruction by the:

A. p53 transcription factor
B. anaphase-promoting complex
C. CDK complex
D. pre-replication complex
E. maturation promoting factor

26. When a female mouse with a defect in a mitochondrial protein required for fatty acid oxidation is crossed with a wild-type male, all progeny (both male and female) have the wild-type phenotype. Which statement is likely correct?

A. Mice do not exhibit maternal inheritance
B. The defect is a result of an autosomal X-linked recessive trait
C. The defect is a result of an X-linked recessive trait
D. The defect is a result of a recessive mitochondrial gene
E. The defect is a result of a nuclear gene mutation

27. The smooth ER participates in:

A. substrate-level phosphorylation
B. exocytosis
C. synthesis of phosphatidylcholine
D. allosteric activation of enzymes
E. synthesis of cytosolic proteins

28. What type of organelle is in plants but not in animals?

A. ribosomes
B. mitochondria
C. nucleus
D. plastids
E. nucleolus

29. Mitochondrial mutations are often limited to one tissue type. If an individual does not produce blood calcium-decreasing hormone calcitonin, which tissue is likely to carry the mitochondrial mutation?

A. thyroid
B. kidney
C. parathyroid
D. liver
E. spleen

30. The rough ER participates in:

A. oxidative phosphorylation
B. synthesis of plasma membrane proteins
C. endocytosis
D. post-translational modification of enzymes
E. synthesis of lipids

31. The best definition of active transport is the movement of:

A. solutes across a semipermeable membrane down an electrochemical gradient
B. solutes across a semipermeable membrane up a concentration gradient
C. substances across a membrane per the Donnan equilibrium
D. solutes via osmosis across a semipermeable membrane from high to low concentration
E. H_2O via diffusion across a semipermeable membrane

32. Overexpression of cyclin D:

A. increases contact inhibition
B. decreases telomerase activity
C. activates apoptosis
D. promotes unscheduled entry into the S phase
E. promotes the transition from G1 to the S phase

33. A failure in which stage of spermatogenesis produces nondisjunction and males with XXY karyotype?

A. prophase I
B. metaphase
C. prophase II
D. telophase
E. anaphase I

34. Protein targeting occurs during the synthesis of which type of proteins?

A. nuclear proteins
B. secreted proteins
C. cytosolic proteins
D. mitochondrial proteins
E. chloroplast proteins

35. The difference between "free" and "attached" ribosomes is that:

I. Free ribosomes are in the cytoplasm, while attached ribosomes are anchored to the endoplasmic reticulum
II. Free ribosomes produce proteins in the cytosol, while attached ribosomes produce proteins that are inserted into the ER lumen
III. Free ribosomes produce proteins that are exported from the cell, while attached ribosomes make proteins for mitochondria and chloroplasts

A. I only
B. II only
C. I and II only
D. I, II and III
E. I and III only

36. The concentration of growth hormone receptors will significantly reduce after selective destruction of which structure?

A. nucleolus
B. nucleus
C. cytosol
D. plasma membrane
E. peroxisomes

37. All these processes are ATP-dependent, EXCEPT:

A. export of Na^+ from a neuron
B. influx of Ca^{2+} into a muscle cell
C. influx of K^+ into a neuron
D. exocytosis of neurotransmitter at a nerve terminus
E. movement of urea across a cell membrane

38. The loss of function of p53 protein results in:

A. blockage in activation of the anaphase-promoting complex
B. increase of contact inhibition
C. elimination of the DNA damage checkpoint
D. activation of apoptosis
E. suppression of spindle fiber assembly

39. Which organelle is most closely associated with exocytosis of newly synthesized secretory protein?

A. peroxisome
B. ribosome
C. lysosome
D. Golgi apparatus
E. nucleus

40. Plant membranes are more fluid than animal membranes because plant membranes:

A. contain substantial amounts of cholesterol
B. have higher amounts of unsaturated fatty acids compared to the membranes of animals
C. have lower amounts of unsaturated fatty acids compared to the membranes of animals
D. are only found in the inner membrane of the mitochondria
E. do not contain glycosylated proteins

41. Which organelles are enclosed in a double membrane?

I. nucleus II. chloroplast III. mitochondrion

A. I and II only
B. II and III only
C. I and III only
D. I, II and III
E. I only

42. Proteins are marked and delivered to specific cell locations through:

A. specific protein transport channels
B. regulation signals released by the cell's cytoskeleton
C. post-translational modifications occurring in the Golgi
D. compartmentalization of the rough ER during protein synthesis
E. post-translational modifications occurring in the nucleus

43. The presence of which element differentiates a protein from a carbohydrate molecule?

A. carbon
B. hydrogen
C. nitrogen
D. oxygen
E. nitrogen and oxygen

44. What change occurs in the capillaries when arterial blood is infused with the plasma protein albumin?

A. Decreased movement of H_2O from the capillaries into the interstitial fluid
B. Increased movement of H_2O from the capillaries into the interstitial fluid
C. Decreased movement of H_2O from the interstitial fluid into the capillaries
D. Increased permeability to albumin
E. Decreased permeability to albumin

45. The retinoblastoma protein controls:

A. contact inhibition
B. transition from G1 to S phase
C. activation of apoptosis
D. expression of cyclin D
E. A and C

46. All processes take place in the mitochondrion, EXCEPT:

A. oxidation of pyruvate
B. Krebs cycle
C. electron transport chain
D. reduction of FADH into $FADH_2$
E. glycolysis

47. Which is the most abundant lipid in the human body?

A. teichoic acid
B. peptidoglycan
C. glycogen
D. triglycerides
E. glucose

48. What is the secretory sequence in the flow of newly synthesized protein for export from the cell?

A. Golgi → rough ER → smooth ER → plasma membrane
B. Golgi → rough ER → plasma membrane
C. smooth ER → rough ER → Golgi → plasma membrane
D. rough ER → smooth ER → Golgi → plasma membrane
E. rough ER → Golgi → plasma membrane

49. Digestive lysosomal hydrolysis would affect all the following EXCEPT:

A. proteins
B. minerals
C. nucleotides
D. lipids
E. carbohydrates

50. Recycling of organelles within the cell is accomplished through autophagy by:

A. mitochondria
B. peroxisomes
C. nucleolus
D. lysosomes
E. rough endoplasmic reticulum

51. All are lipid derivatives EXCEPT:

A. carotenoids
B. albumins
C. waxes
D. steroids
E. lecithin

52. Defective attachment of a chromosome to the spindle:

A. blocks activation of the anaphase-promoting complex
B. activates exit from mitosis
C. prevents overexpression of cyclin D
D. activates sister chromatid separation
E. promotes the transition from G1 to the S phase

53. Which stage is when human cells with a single unreplicated copy of the genome are formed?

A. mitosis
B. meiosis II
C. meiosis I
D. interphase
E. G1

54. Which organelle in the cell is the site of fatty acid, phospholipid, and steroid synthesis?

A. endosome
B. peroxisome
C. rough endoplasmic reticulum
D. smooth endoplasmic reticulum
E. chloroplast

55. Which eukaryotic organelle is NOT membrane-bound?

A. nucleus
B. plastid
C. centriole
D. chloroplast
E. C and D

56. Which phase of mitotic division do spindle fibers split the centromere and separate the sister chromatids?

A. interphase
B. telophase
C. prophase
D. metaphase
E. anaphase

57. All are correct about cyclic AMP (cAMP), EXCEPT:

A. the enzyme that catalyzes the formation of cAMP is in the cytoplasm
B. membrane receptors can activate the enzyme that forms cAMP
C. ATP is the precursor molecule in the formation of cAMP
D. adenylate cyclase is the enzyme that catalyzes the formation of cAMP
E. cAMP is a second messenger that triggers a cascade of intracellular reactions when a peptide hormone binds to a receptor on the cell membrane

58. Which molecule, and its associated protein kinase, ensures a proper progression of cell division?

A. oncogene
B. tumor suppressor
C. cyclin
D. histone
E. homeotic

59. Placed in a hypertonic solution, erythrocytes will undergo:

A. crenation
B. expansion
C. plasmolysis and rupture
D. no change
E. shrinkage and then rapid expansion

60. Tiny organelles abundant in the liver contain oxidases and detoxify substances (i.e., alcohol and hydrogen peroxide) are:

A. centrosome
B. lysosomes
C. endosomes
D. rough endoplasmic reticulum
E. peroxisomes

61. Microtubules can function independently or form protein complexes to produce structures like:

I. flagella
II. actin and myosin filaments in muscle cells
III. mitotic spindle apparatus

A. III only
B. I and III only
C. II and III only
D. I and II only
E. I, II, and III

62. About a cell that secretes much protein (e.g., a pancreatic exocrine cell), it can be assumed that this cell has:

I. an abundance of rough endoplasmic reticulum
II. a large Golgi apparatus
III. a prominent nucleolus

A. I only
B. I and III only
C. II and III only
D. I, II and III
E. I and II only

63. A cell division where each daughter cell receives a chromosome complement identical to the parent is:

A. mitosis
B. non-disjunction
C. meiosis
D. replication
E. crossing-over

64. Which is affected LEAST by colchicine, which interferes with microtubule formation?

A. organelle movement
B. meiosis
C. mitosis
D. cilia
E. pseudopodia for amoeboid motility

65. Which process is NOT an example of apoptosis?

A. Reabsorption of a tadpole's tail during metamorphosis into a frog
B. Formation of the endometrial lining of the uterus during the menstrual cycle
C. Formation of the synaptic cleft by triggering cell death in brain neuronal cells
D. Formation of fingers in the fetus by the removal of tissue between the digits
E. All are examples of apoptosis

66. Cells that utilize large quantities of ATP for active transport (e.g., epithelial cells of the intestine) have:

A. many mitochondria
B. elevated levels of DNA synthesis
C. high levels of adenylate cyclase
D. polyribosomes
E. many lysosomes

67. Clathrin is a protein collected on the cytoplasmic side of cell membranes and functions in the coordinated pinching off membrane into receptor-mediated endocytosis. It can be predicted that a lipid-soluble toxin that inactivates clathrin results in:

A. increased ATP consumption
B. increased protein production on the rough endoplasmic reticulum
C. increased secretion of hormone into the extracellular fluid
D. increased ATP production
E. reduced delivery of polypeptide hormones to endosomes

68. Cells that respond to peptide hormones usually do so through biochemical reactions involving membrane receptors and kinase activation. For cells to respond, the first and second messengers communicate because:

A. peptide hormones pass through the cell membrane and elicit a response
B. the hormone-receptor complex moves into the cytoplasm as a unit
C. hormones alter cellular activities directly through gene expression
D. the G protein acts as a link between the first and second messengers
E. the presence of a cytosolic receptor binds the hormone

69. Mitosis does NOT serve the purpose of:

A. replenishment of erythrocytes
B. formation of scar tissue
C. organ repair
D. tissue growth
E. transduction

Notes for active learning

Cellular Metabolism & Enzymes

1. The atom that generates a hydrogen bond to stabilize the α-helical configuration of a polypeptide is:

A. peptide bond atom
B. atom in the R-groups
C. hydrogen of carbonyl oxygen
D. hydrogen of the amino nitrogen
E. two of the above

2. The ATP molecule contains three phosphate groups, two of which are:

A. bound as phosphoanhydrides
B. bound to adenosine
C. never hydrolyzed from the molecule
D. cleaved off during most biochemical reactions
E. equivalent in energy for the hydrolysis of each of the phosphates

3. Which attractive force is used by side chains of nonpolar amino acids to interact with nonpolar amino acids?

A. ionic bonds
B. hydrogen bonds
C. hydrophobic interaction
D. disulfide bonds
E. dipole-dipole

4. Fermentation yields less energy than aerobic respiration because:

A. it requires a greater expenditure of cellular energy
B. glucose molecules are not completely oxidized
C. it requires more time for ATP production
D. oxaloacetic acid serves as the final H^+ acceptor
E. it occurs in H_2O

5. How would the reaction kinetics of an enzyme and its substrate change if an anti-substrate antibody is added?

A. The antibody binds to the substrate, which increases the V_{max}
B. The antibody binds to the substrate, which decreases K_m
C. No change because K_m and V_{max} are independent of antibody concentration
D. The antibody binds the substrate, which decreases V_{max}
E. The antibody binds to the substrate, which increases K_m

6. Metabolism is:

A. consumption of energy
B. release of energy
C. all conversions of matter and energy taking place in an organism
D. production of heat by chemical reactions
E. exchange of nutrients and waste products with the environment

7. During alcoholic fermentation, all occur EXCEPT:

A. release of CO_2
B. oxidation of glyceraldehyde-3-phosphate
C. oxygen is not consumed in the reaction
D. ATP synthesis as a result of oxidative phosphorylation
E. NAD^+ is produced

8. When determining a protein's amino acid sequence, acid hydrolysis causes partial destruction of tryptophan, conversion of asparagine into aspartic acid, and conversion of glutamine into glutamic acid. Which statement is NOT correct?

A. Glutamine concentration is related to the level of aspartic acid
B. Glutamic acid levels are an indirect indicator of glutamine concentration
C. Tryptophan levels cannot be estimated accurately
D. Asparagine levels cannot be estimated accurately
E. Tryptophan and asparagine levels cannot be estimated accurately

9. What is the correct sequence of energy sources used by the body?

A. fats → glucose → other carbohydrates → proteins
B. glucose → other carbohydrates → fats → proteins
C. glucose → other carbohydrates → proteins → fats
D. glucose → fats → proteins → other carbohydrates
E. fats → proteins → glucose → other carbohydrates

10. Enzymes act by:

A. lowering the overall free energy change of the reaction
B. decreasing the distance reactants must diffuse to find each other
C. increasing the activation energy
D. shifting equilibrium towards product formation
E. decreasing the activation energy

11. For the following reaction, which statement is TRUE?

ATP + Glucose → Glucose-6-phosphate + ADP

A. reaction results in the formation of a phosphodiester bond
B. reaction is endergonic
C. reaction is part of the Krebs cycle
D. free energy change for the reaction is approx. –4 kcal
E. reaction does not require an enzyme

12. Which is the correct cAMP classification because cAMP-dependent protein phosphorylation activates hormone-sensitive lipase?

A. DNA polymerase
B. lipoproteins
C. glycosphingolipids
D. second messenger
E. phospholipids

13. Which statement is NOT true about the Krebs cycle?

A. Krebs cycle occurs in the matrix of the mitochondria
B. Citrate is an intermediate in the Krebs cycle
C. Krebs cycle produces nucleotides such as NADH and $FADH_2$
D. Krebs cycle is linked to glycolysis by pyruvate
E. Krebs cycle is the single greatest direct source of ATP in the cell

14. The rate of V_{max} is directly related to:

I. Enzyme concentration
II. Substrate concentration
III. Concentration of a competitive inhibitor

A. I, II, and III
B. I and III only
C. I and II only
D. I only
E. II and III only

15. A reaction in which the substrate glucose binds to the enzyme hexokinase, and the conformation of both molecules changes, is an example of:

A. lock-and-key mechanism
B. induced-fit mechanism
D. allosteric inhibition
E. covalent bond formation at the active site
C. competitive inhibition

16. You are studying an enzyme that catalyzes a reaction with a free energy change of +5 kcal. If you double the amount of enzyme in a reaction mixture, what would be the free energy change for the reaction?

A. –10 kcal
B. –5 kcal
C. 0 kcal
D. +5 kcal
E. +10 kcal

17. Glucokinase and hexokinase catalyze the first glycolysis reaction; glucokinase has a higher K_m. Which is a correct statement if K_m is equal to [Substrate] = $1/2V_{max}$?

A. hexokinase is always functional and is not regulated by negative feedback
B. hexokinase and glucokinase are not isozymes
C. glucokinase is not a zymogen
D. glucokinase becomes active from elevated levels of fructose
E. none of the above

18. α-helices and β-pleated sheets are characteristic of which level of protein folding?

A. primary
B. secondary
C. tertiary
D. quaternary
E. secondary & tertiary

19. Coenzymes are:

A. minerals such as Ca^{2+} and Mg^{2+}
B. small inorganic molecules that work with an enzyme to enhance the reaction rate
C. linking together of two or more enzymes
D. small molecules that do not regulate enzymes
E. small organic molecules that work with an enzyme to enhance the reaction rate

20. Hemoglobin is an example of a protein that:

A. is initially inactive in cell
B. has a quaternary structure
C. conducts a catalytic reaction
D. has only a tertiary structure
E. has a signal sequence

21. The site of the TCA cycle in eukaryotic cells, as opposed to prokaryotes, is:

A. mitochondria
B. endoplasmic reticulum
C. cytosol
D. nucleolus
E. intermembrane space of the mitochondria

22. All are metabolic waste products, EXCEPT:

A. lactate
B. pyruvate
C. CO_2
D. H_2O
E. ammonia

23. When measuring the reaction velocity as a function of substrate concentration, what is likely to occur if the enzyme concentration changes?

A. V_{max} changes, while K_m remains constant
B. V_{max} remains constant, while V changes
C. V_{max} remains constant, while K_m changes
D. V_{max} remains constant, while V and K_m change
E. Not possible to predict without experimental data

24. A holoenzyme is:

A. inactive enzyme without its cofactor
B. inactive enzyme without its coenzyme
C. active enzyme with its organic moiety
D. active enzyme with its coenzyme
E. active enzyme with its cofactor

25. *Clostridium butyricum* is a heterotrophic anaerobe that grows on glucose, converting it to butyric acid as a product. If the free energy for this reaction is –50 kcal, the maximum number of ATP that this organism can synthesize from one molecule of glucose is approximately:

A. 5 ATP
B. 36 ATP
C. 7 ATP
D. 38 ATP
E. 0 ATP

26. Which amino acid is directly affected by dithiothreitol (DTT), known to reduce and break disulfide bonds?

A. methionine
B. leucine
C. glutamine
D. cysteine
E. proline

27. Which answer represents a correct pairing of aspects for cellular respiration?

A. Krebs cycle – cytoplasm
B. fatty acid degradation – lysosomes
C. electron transport chain – inner mitochondrial membrane
D. glycolysis – inner mitochondrial membrane
E. ATP synthesis – outer mitochondrial membrane

28. An apoenzyme is:

A. active enzyme with its organic moiety
B. inactive enzyme without its inorganic cofactor
C. active enzyme with its cofactor
D. active enzyme with its coenzyme
E. inactive enzyme without its cofactor

29. Several forces stabilize tertiary structure of a protein. Which is likely involved in this stabilization?

A. glycosidic bonds
B. disulfide bonds
C. peptide bonds
D. anhydride bonds
E. phosphodiester bonds

30. In the non-oxidative branch of the pentose phosphate pathway, transketolase is a reaction catalyst enzyme, and its activity depends on a prosthetic group. Which bond is used by a prosthetic group to attach to its target?

A. van der Waals interactions
B. covalent bond
C. ionic bond
D. hydrogen bond
E. dipole-dipole interactions

31. The process of $C_6H_{12}O_6 + O_2 \rightarrow CO_2 + H_2O$ is completed in the:

A. plasma membrane
B. cytoplasm
C. ribosome
D. nucleus
E. mitochondria

32. In hyperthyroidism, oxidative metabolism rates measured through basal metabolic rate (BMR) will be:

A. indeterminable
B. below normal
C. above normal
D. normal
E. between below normal to normal

33. Cofactors are:

I. small inorganic molecules that work with an enzyme to enhance the reaction rate
II. small organic molecules that work with an enzyme to enhance the reaction rate
III. small molecules that regulate enzyme activity

A. I only
B. I and II only
C. II and III only
D. I, II and III
E. I and III only

34. All proteins:

A. are post-translationally modified
B. have a primary structure
C. have catalytic activity
D. contain prosthetic groups
E. contain disulfide bonds

35. After being gently denatured with the denaturant removed, proteins can recover significant activity because recovery of the structure depends on?

A. 4° structure of the polypeptide
B. 3° structure of the polypeptide
C. 2° structure of the polypeptide
D. 1° structure of the polypeptide
E. interactions between the polypeptide and its prosthetic groups

36. All statements about glycolysis are true, EXCEPT:

A. end-product can be lactate, ethanol, CO_2, and pyruvate
B. $FADH_2$ is produced during glycolysis
C. a molecule of glucose is converted into two molecules of pyruvate
D. net total of two ATPs are produced
E. NADH is produced

37. Vitamins are:

A. necessary components in the human diet
B. present in plants but not in animals
C. absent in bacteria within the gastrointestinal tract
D. all water-soluble
E. inorganic components of the diet

38. Hemoglobin is a protein that contains a:

A. site where proteolysis occurs
B. phosphate group at its active site
C. bound zinc atom
D. prosthetic group
E. serine phosphate at its active site

39. Which amino acid is nonoptically active, lacking four different groups bonded to the α carbon?

A. valine
B. aspartic acid
C. glutamate
D. cysteine
E. glycine

40. Which statement is TRUE for the glycolytic pathway?

A. glucose produces a net of two molecules of ATP and two molecules of NADH
B. glucose produces one molecule of pyruvate
C. O_2 is a reactant for glycolysis
D. glucose is partially reduced
E. pyruvate is the final product of the Krebs cycle and is intermediate for the next series of reactions in cellular respiration

41. Which metabolic process occurs in the mitochondria?

I. Krebs cycle   II. glycolysis   III. electron transport chain

A. II only
B. I and III only
C. II and III only
D. I, II and III
E. I and II only

42. Which statement below best describes the usual relationship of the inhibitor molecule to the allosteric enzyme in feedback inhibition of enzyme activity?

A. The inhibitor is the substrate of the enzyme
B. The inhibitor is the product of the enzyme-catalyzed reaction
C. The inhibitor is the final product of the metabolic pathway
D. The inhibitor is a metabolically unrelated signal molecule
E. The inhibitor binds to a tertiary protein

43. *Clostridium butyricum* is an obligate anaerobe that grows on glucose and converts it to butyric acid. If the ΔG for this reaction is –50 kcal, the synthesis of ATP occurs through:

A. substrate-level phosphorylation
B. oxidative phosphorylation
C. neither substrate-level nor oxidative phosphorylation
D. both substrate-level and oxidative phosphorylation
E. electron transport cascade

44. While covalent bonds are the strongest bonds of protein structure, which are connected by a peptide bond?

A. ammonium group and ester group
B. two amino groups
C. the α carbons
D. two carboxylate groups
E. amino group and carboxylate group

45. All statements apply to oxidative phosphorylation, EXCEPT:

A. it can occur under anaerobic conditions
B. it produces two ATPs for each $FADH_2$
C. it involves O_2 as the final electron acceptor
D. it occurs on the inner membrane of the mitochondrion
E. it involves a cytochrome electron transport chain

46. Like other catalysts, enzymes:

I. increase the rate of reactions without affecting ΔG
II. shift the chemical equilibrium from more reactants to more products
III. do not alter the chemical equilibrium between reactants and products

A. I only
B. I and II only
C. I and III only
D. III only
E. II and III only

47. Enzyme activity can be regulated by:

I. zymogen proteolysis
II. changes in substrate concentration
III. post-translational modifications

A. I only
B. II and III only
C. I, II and III
D. I and II only
E. I and III only

48. Which interactions stabilize parallel and non-parallel beta-pleated sheets?

A. hydrophobic interactions
B. hydrogen bonds
C. van der Waals interactions
D. covalent bonds
E. dipole-dipole interactions

49. Glycogen is:

A. degraded by glycogenesis
B. synthesized by glycogenolysis
C. unbranched molecule
D. in both plants and animals
E. the storage polymer of glucose

50. If [S] = 2 K_m, what portion of active sites of the enzyme is filled by substrate?

A. 3/4
B. 2/3
C. 1/2
D. 1/3
E. 1/4

51. In allosteric regulation, how is enzyme activity affected by binding a small regulatory molecule to the enzyme?

A. It is inhibited
B. It is stimulated
C. Can be either stimulated or inhibited
D. Is neither stimulated nor inhibited
E. The rate increases to twice K_m and then plateaus

52. All biological reactions:

A. are exergonic
B. have an activation energy
C. are endergonic
D. occur without a catalyst
E. are irreversible

53. In eukaryotes, energy is trapped in a high-energy phosphate group during oxidative phosphorylation in the:

A. nucleus
B. mitochondrial matrix
C. inner mitochondrial membrane
D. outer mitochondrial membrane
E. cytoplasmic face of the plasma membrane

54. All statements about enzymes are true, EXCEPT:

A. They function optimally at a particular temperature
B. They function optimally at a particular pH
C. They may interact with non-protein molecules to achieve biological activity
D. Their activity is not affected by a genetic mutation
E. They are almost always proteins

55. The Gibbs free-energy change (ΔG) of a reaction is determined by:

I. intrinsic properties of the reactants and products
II. concentrations of the reactants and products
III. the temperature of the reactants and products

A. I only
B. I and III only
C. II and III only
D. I, II and III
E. I and II only

56. Allosteric enzymes:

I. are regulated by metabolites that bind at sites other than the active site
II. have quaternary structure
III. show cooperative binding of substrate

A. I only
B. I and II only
C. II and III only
D. I, II and III
E. III only

57. Which symptom is characteristic of a patient exposed to monoamine oxidase inhibitors, given that they prevent the breakdown of catecholamines (i.e., epinephrine)?

A. decreased blood flow to skeletal muscles
B. excessive digestive activity
C. dilated pupils
D. decreased heart rate
E. increased peristalsis along the GI tract

58. The ΔG for hydrolysis of ATP to ADP and Pi is:

A. Greater than +7.3 kcal/mole
B. +7.3 kcal/mole
C. –7.3 kcal/mole
D. –0.5 kcal/mole
E. none of the above

59. The active site of an enzyme is where:

I. prosthetic group is bound
II. proteolysis occurs for zymogens
III. non-competitive inhibitors bind

A. I only
B. II only
C. I and II only
D. II and III only
E. I, II and III

60. The hydrolysis of ATP → ADP + phosphate allows glucose-6-phosphate to be synthesized from glucose and phosphate because:

A. heat produced from ATP hydrolysis drives the glucose-6-phosphate synthesis
B. enzymatic coupling of these two reactions allows the energy of ATP hydrolysis to drive the synthesis of glucose-6-phosphate
C. energy of glucose phosphorylation drives ATP splitting
D. all the above
E. None of the above

61. Which statement is TRUE?

A. Protein function can be altered by modification after synthesis
B. Covalent bonds stabilize the secondary structure of proteins
C. A protein with a single polypeptide subunit has a quaternary structure
D. Integral membrane proteins contain high amounts of acidic amino acids
E. Protein denaturation is always reversible

62. What does alcoholic fermentation have in common with pyruvate oxidation under aerobic conditions?

A. no commonality
B. triose sugar is a product of each reaction
C. ethyl alcohol is a product of each reaction
D. CO_2 is a product of each reaction
E. NADH is a product of each reaction

63. Which statement is TRUE?

A. All disaccharides must contain fructose
B. Most polysaccharides contain ribose
C. All polysaccharides are energy-generating molecules
D. Polysaccharides are only in animal cells
E. Glucose and fructose have different chemical properties even with the same molecular formula

Notes for active learning

Notes for active learning

Photosynthesis

1. What mechanism is used by C_4 plants to conserve water?

   A. developing deep roots
   B. performing the Calvin cycle during the day
   C. developing water storage in their leaves and stems
   D. performing the Calvin cycle early in the morning
   E. closing the stomata during heat and dryness

2. What mechanism is used by CAM plants to conserve water?

   A. performing the Calvin cycle during the day
   B. developing water storage in their leaves and stems
   C. including carbon dioxide into RuBP
   D. opening their stomata during the night
   E. closing their stomata during the night

3. Which of the choices is an autotroph?

   A. beech tree
   B. octopus
   C. zooplankton
   D. coral
   E. dog

4. Which is NOT a true statement about ATP?

   A. ATP provides energy for the mechanical functions of cells
   B. Used ATP is discarded by the cell as waste
   C. ATP consists of ribose, adenine, and three phosphate groups
   D. ADP is produced when ATP releases energy
   E. ATP conversion to ADP is an exothermic reaction

5. Photosynthesis has a role in the metabolism of a plant through:

   A. breaking down sugars into H_2O and O_2
   B. converting H_2O into CO_2
   C. consuming CO_2 and synthesizing sugars
   D. converting O_2 into cellulose
   E. converting O_2 into sugars

6. From the figure, all are parts of an ADP molecule, EXCEPT:

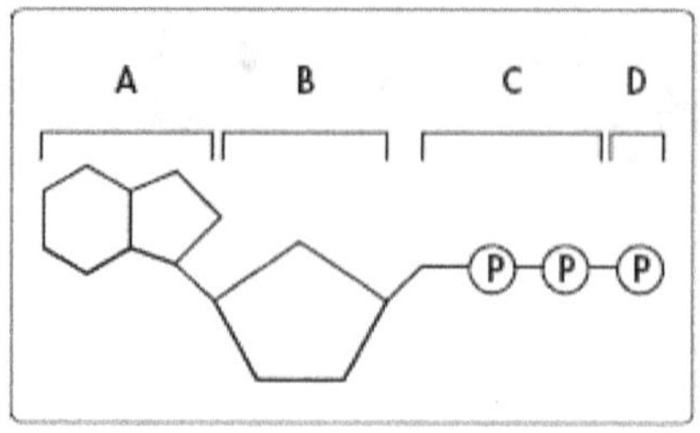

A. structure A
B. structure B
C. structure C
D. structure D
E. structures A and B

7. Within the inner membrane of a chloroplast, stacks of thylakoids are surrounded by:

A. stroma fluid
B. grana
C. chloroplast
D. thylakoids
E. cristae

8. What happens during photosynthesis?

A. Autotrophs consume carbohydrates
B. Autotrophs produce carbohydrates
C. Heterotrophs consume ATP
D. Heterotrophs produce ATP
E. Heterotrophs produce sugars

9. The role of stomata is to facilitate:

A. gas exchange
B. H_2O release
C. Calvin cycle
D. H_2O uptake
E. Chloroplast formation

10. Plants gather the sun's energy using:

A. glucose
B. chloroplasts
C. thylakoids
D. mitochondria
E. pigments

11. Which structures are used by plants to obtain most of their water?

A. Stomata
B. Flowers
C. Roots
D. Chloroplasts
E. Shoots

12. Which structure in the image illustrates a single thylakoid?

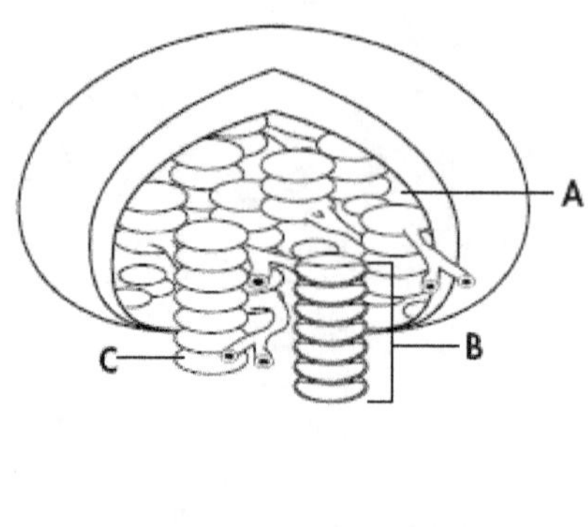

A. Structure A
B. Structure B
C. Structure C
D. Structures A and B
E. Structures B and C

13. Membranous structures in chloroplasts are:

A. organelles
B. grana
C. stomata
D. plasma
E. thylakoids

14. Chlorophyll is within chloroplast in the:

A. thylakoid space
B. thylakoid membrane
C. stroma
D. ATP
E. space between the inner and outer membrane of the thylakoid

15. Grana is/are:

A. stacks of membranous sacs
B. chloroplast pigments
C. fluid in the chloroplasts
D. the space between the inner and outer membrane of chloroplasts
E. the membrane-bound compartment of chloroplasts

16. Which chemical shown in the figure is an electron carrier molecule?

A. $NADP^+$
B. Oxygen
C. H_2O
D. CO_2
E. Sugars

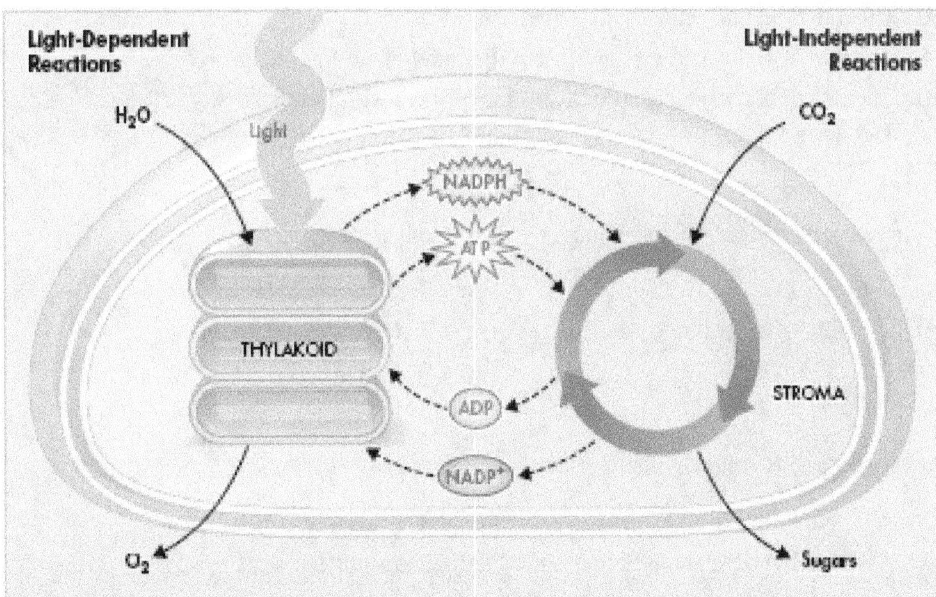

17. Which is likely to result if a shade-tolerant plant receives a minimal amount of water while receiving its necessary amount of light?

A. Accelerated plant growth
B. Low ATP production
C. Increased output of oxygen
D. Increased consumption of carbon dioxide
E. Plant death

18. Why are electron carriers needed for transporting electrons from one part of the chloroplast to another?

A. High-energy electrons get their energy from electron carriers
B. High-energy electrons are not soluble in the cytoplasm
C. High-energy electrons are easily oxidized
D. High-energy electrons would be destroyed
E. High-energy electrons are highly reactive

19. Which equation best describes the reaction of photosynthesis?

A. $6\ O_2 + 6\ CO_2 \rightarrow C_6H_{12}O_6 + 6\ H_2O$
B. $C_6H_{12}O_6 + 6\ H_2O \rightarrow 6\ CO_2 + 6\ O_2$
C. $C_6H_{12}O_6 \rightarrow 6\ CO_2 + 6\ H_2O + 6\ O_2$
D. $6\ H_2O + 6\ CO_2 \rightarrow C_6H_{12}O_6 + 6\ O_2$
E. $6\ CO_2 + 6\ O_2 + 6\ H_2O \rightarrow C_6H_{12}O_6$

20. What is the result of an increase in light intensity that a plant receives?

A. The rate of photosynthesis increases and then levels off
B. The rate of photosynthesis does not change
C. The rate of photosynthesis increases indefinitely with light intensity
D. The rate of photosynthesis decreases indefinitely with light intensity
E. The rate of photosynthesis increases indefinitely to the square root of the light intensity

21. In redox reactions of photosynthesis, electron transfer goes as follows:

A. $H_2O \rightarrow CO_2$
B. $C_6H_{12}O_6 \rightarrow O_2$
C. $O_2 \rightarrow C_6H_{12}O_6$
D. $CO_2 \rightarrow H_2O$
E. $C_6H_{12}O_6 \rightarrow H_2O$

22. Which is NOT a step in the light-dependent reactions?

A. ATP synthase allows H^+ ions to pass through the thylakoid membrane
B. ATP and NADPH are used to produce high-energy sugars
C. High-energy electrons move through the electron transport chain
D. Pigments in photosystem II absorb light
E. Electrons move from photosystem II to photosystem I

23. Which is the product of the light reaction of photosynthesis?

A. sugar
B. carbon monoxide
C. water
D. carbon dioxide
E. oxygen

24. Photosystems I and II are in the:

A. cell membrane
B. Calvin cycle
C. thylakoid membrane
D. stroma
E. matrix of the mitochondrion

25. Which components from the light reactions are required by the C_3 cycle?

A. NADH and RuBP
B. NADPH and ATP
C. ATP and NADH
D. glucose and $NADP^+$
E. water and $NADP^+$

26. Which pathway is the correct flow of electrons during photosynthesis?

A. Photosystem I → Calvin cycle → $NADP^+$
B. Light → Photosystem I → Photosystem II
C. H_2O → Photosystem I → Photosystem II
D. O_2 → ADP → Calvin cycle
E. H_2O → $NADP^+$ → Calvin cycle

27. Light-dependent reactions of photosynthesis produce:

A. H_2O and RuBP
B. ATP and CO_2
C. NADH and ATP
D. NADPH and ATP
E. H_2O and O_2

28. The Calvin cycle occurs in the:

A. chlorophyll molecules
B. thylakoid membranes
C. photosystems
D. stroma
E. matrix

29. A wavelength of which light is reflected by a blue-colored plant?

A. violet
B. yellow
C. blue
D. lime
E. purple

30. What are the three parts of an ATP molecule?

A. Adenine, ribose, and three phosphate groups
B. NADH, NADPH, and $FADH_2$
C. Adenine, thylakoid, and a phosphate group
D. Stroma, grana, and chlorophyll
E. Adenine, ribose, and a phosphate group

31. When observing a plant, its visible color is:

A. wavelength emitted by the plant
B. wavelength from the return of excited electrons to their ground state
C. wavelength being reflected by that plant
D. wavelength absorbed by the pigment of the plant
E. wavelength from the excited electrons of the plant

32. Energy is released from ATP when:

A. a phosphate group is removed
B. ATP is exposed to sunlight
C. adenine binds to ribose
D. a phosphate group is added
E. a base is removed

33. The wavelength energy least utilized by photosynthesis is:

A. orange
B. blue
C. violet
D. cyan
E. green

34. Organisms, such as plants, which make their food are:

A. symbiotic
B. parasitic
C. autotrophs
D. heterotrophs
E. omnivores

35. During the fall foliage, which pigments are responsible for the yellow, red, and orange colors of leaves?

A. carotenoids
B. chlorophyll *b*
C. melanin
D. chlorophyll *a*
E. anthocyanin

36. Which organism is a heterotroph?

A. sunflower
B. flowering plant
C. alga
D. mushroom
E. deciduous trees

37. In addition to chlorophyll *a*, plants have chlorophyll *b* and carotenoids accessory pigments because:

A. there is not enough chlorophyll *a* produced for the plant's energy needs
B. plants must have leaves of different colors
C. these pigments reflect more energy
D. these pigments protect plants from UV radiation
E. these pigments absorb energy wavelengths that chlorophyll *a* does not

38. Plants get the energy they need for photosynthesis by absorbing:

A. chlorophyll *b*
B. energy from the sun
C. high-energy sugars
D. chlorophyll *a*
E. both chlorophyll *a* and *b*

39. A discrete packet of light is a(n):

A. photon
B. neutron
C. proton
D. wavelength
E. electron

40. Most plants appear green because chlorophyll:

A. does not absorb violet light
B. does not absorb green light
C. absorbs violet light
D. absorbs green light
E. generate wavelengths of green light

41. In visible light, shorter wavelengths carry:

A. more energy
B. more photons
C. more red color
D. less energy
E. fewer photons

42. Interconnected sacs of a membrane suspended in a thick fluid are:

A. stroma
B. thylakoids
C. chlorophyll
D. grana
E. stomata

43. When a molecule absorbs a photon, one of its electrons is raised to a(n):

A. higher state
B. excited state
C. lower state
D. ground state
E. valence

44. What is the function of $NADP^+$ in photosynthesis?

A. Photosystem
B. Pigment
C. Electron carrier
D. High-energy sugar
E. Absorb photons

45. A molecule releases energy gained from the absorption of a photon through:

I. fluorescence II. heat III. light

A. I only
B. II only
C. III only
D. II and III only
E. I, II and III

46. Photosynthesis uses sunlight to convert water and carbon dioxide into:

A. oxygen and high-energy sugars
B. ATP and oxygen
C. high-energy sugars and proteins
D. oxygen and carbon
E. ATP and high-energy sugars

47. Photosystems are in:

A. grana
B. stroma
C. stomata
D. thylakoid membranes
E. matrix of mitochondria

48. The light-dependent reactions take place:

A. within the thylakoid membranes
B. in the outer membrane of the chloroplast
C. in the stroma of the chloroplast
D. within the mitochondria membranes
E. in the inner membrane of the chloroplast

49. Which compound is in the photosystem's reaction center?

A. FADPH
B. rhodopsin
C. ATP
D. ADP
E. chlorophyll *a*

50. What are the products of the light-dependent reactions?

A. CO_2 gas, O_2 gas, and NADPH
B. ATP, CO_2 gas, and NADPH
C. ATP, NADPH, and O_2 gas
D. O_2 gas and glucose
E. ATP, NADPH, and glucose

51. Which is the correct matching of molecules and products in the Calvin cycle?

A. ATP + NADPH + carbon dioxide ⇒ sugar
B. light + water + carbon dioxide ⇒ sugar and oxygen
C. ATP + NADPH + carbon dioxide ⇒ sugar and oxygen
D. light + water + carbon dioxide ⇒ sugar
E. carbon dioxide + light ⇒ oxygen and water

52. Which activities happen within the stroma?

A. Electrons move through the electron transport chain
B. The Calvin cycle produces sugars
C. ATP synthase produces ATP
D. Photosystem I absorbs light
E. Photosystem II produces sugar

53. Oxygen released by a photosystem comes from:

A. ATP
B. carbon dioxide
C. Chlorophyll *a*
D. water
E. light

54. The Calvin cycle is another name for the:

A. photosynthesis reaction
B. electron transport chain
C. light-independent reactions
D. light-dependent reactions
E. Photosystem I

55. Electrons for light reactions originate from:

A. NADPH
B. water
C. light
D. carbon dioxide
E. ATP

56. Which mechanism occurs in light reactions of photosynthesis and cellular respiration?

A. C_3 cycle
B. Beta oxidation
C. Glycolysis
D. Krebs cycle
E. Electron transport chain

57. During photosynthesis, an H^+ ion gradient is formed across:

A. mitochondrial inner membrane
B. inner chloroplast membrane
C. thylakoid membrane
D. mitochondrial outer membrane
E. stroma membrane

Notes for active learning

Notes for active learning

Section II: Molecular Biology of Cells

Molecular Biology of Eukaryotes

1. Which primers should be used to amplify the DNA shown in the figure below via PCR?

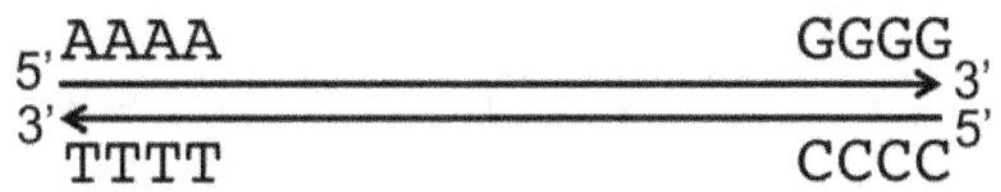

A. 5'-CCCC-3' and 5'-AAAA-3'
B. 5'-GGGG-3' and 5'-TTTT-3'
C. 5'-AAAA-3' and 5'-GGGG-3'
D. 5'-TTTT-3' and 5'-CCCC-3'
E. none of the above

2. The major phenotypic expression of a genotype is in:

A. rRNA
B. tRNA
C. mRNA
D. nucleic acids
E. proteins

3. Which molecule is used in DNA sequencing to cause termination when the template strand is G?

I.

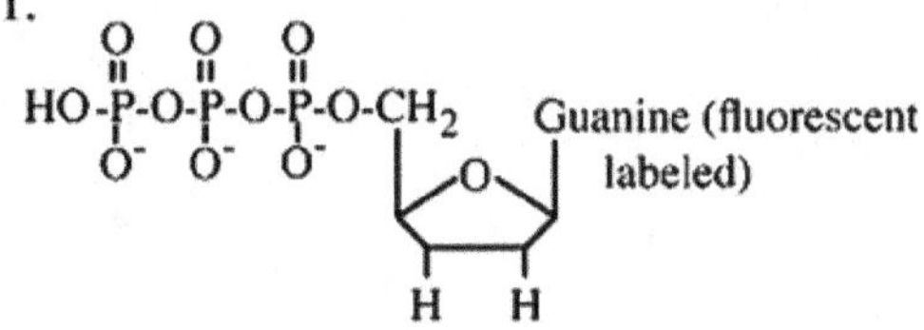

III.

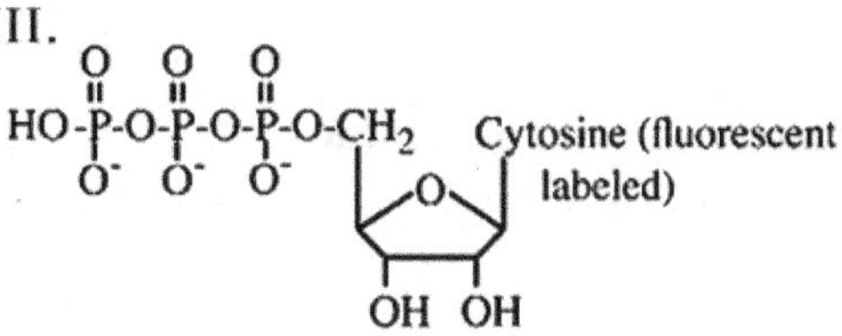

II.

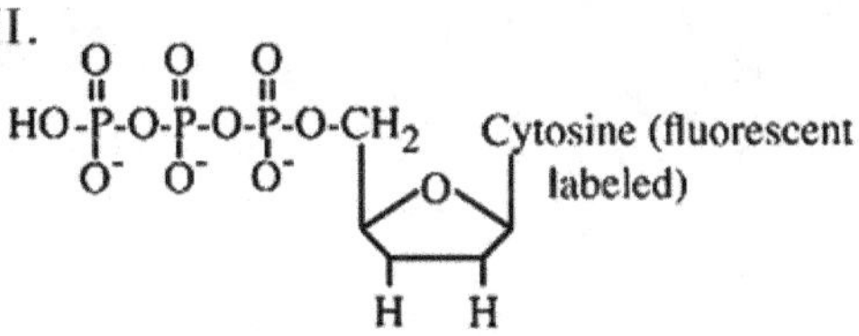

IV.

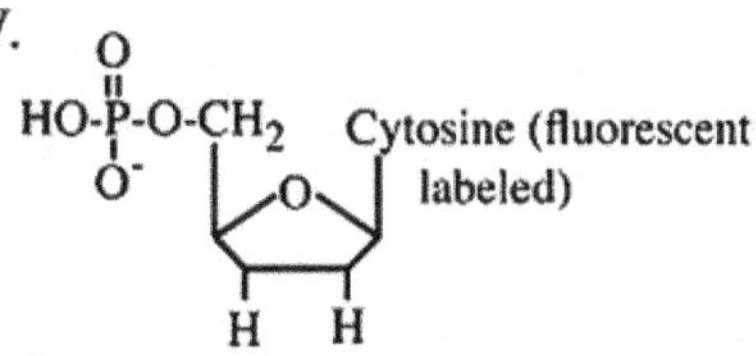

A. molecule I & III
B. molecule III & IV
C. molecule II
D. molecule IV
E. molecule III

4. Which is NOT a part of the post-translational modification of protein?

A. addition of a 3' poly-A tail
B. phosphorylation
C. methylation
D. glycosylation
E. acetylation

5. A chromosome with its centromere in the middle is:

A. acrocentric
B. telocentric
C. metacentric
D. holocentric
E. midcentromeric

6. Which is NOT an example of an environmental factor affecting how a gene is expressed?

A. Heat shock proteins are synthesized in cells after a temperature increase
B. *Drosophila* with specific genes develop bent wings when incubated at low temperatures and straight wings when incubated at elevated temperatures
C. Himalayan hares change hair color after cooling in the naturally warm regions
D. Shivering occurs after a decrease in body temperature
E. All the above

7. cDNA libraries contain:

A. promoters
B. intron portions of expressed genes
C. exon portions of expressed genes
D. non-expressed retrotransposons
E. both introns and exons of expressed genes

8. Individual genes often encode for:

I. enzymes with tertiary structure
II. enzymes with quaternary structure
III. complex polysaccharides

A. I only
B. II only
C. I and II only
D. I, II and III
E. I and III only

9. What enzyme is often used to make a genomic library?

A. RNA polymerase
B. reverse transcriptase
C. deoxyribonuclease
D. DNA polymerase
E. restriction endonuclease

10. Attachment of glycoprotein side chains and amino acid hydroxylation are post-translational modifications. What is the site for protein glycosylation?

I. Lysosomes
II. Golgi apparatus
III. Rough endoplasmic reticulum

A. II only
B. III only
C. I and II only
D. II and III only
E. I, II and III

11. The enzyme used for restoring the ends of the DNA in a chromosome is:

A. telomerase
B. helicase
C. polymerase
D. gyrase
E. ligase

12. A cDNA library is made using:

A. DNA from the region where the gene of interest is expressed
B. mRNA from the region where the gene of interest is expressed
C. mRNA from the region where the gene of interest is not expressed
D. rRNA from the region where the gene of interest is expressed
E. all the above

13. Which statement is the *central dogma* of molecular biology?

A. Information flow between DNA, RNA, and protein is reversible
B. Information flow in the cell is unidirectional, from protein to RNA to DNA
C. The genetic code is ambiguous but not degenerate
D. The DNA sequence of a gene can be predicted from the amino acid sequence of the protein
E. Information flow in the cell is unidirectional, from DNA to RNA to protein

14. What is alternative splicing?

A. Cleavage of peptide bonds to create different proteins
B. Cleavage of DNA to make different genes
C. Cleavage of hnRNA to make different mRNAs
D. A new method of splicing that does not involve snRNPs
E. None of the above

15. This compound is synthesized by the nucleolus and is necessary for ribosomal function.

A. ribozyme
B. riboflavin
C. liposome
D. rRNA
E. tRNA

16. *E. coli* RNA polymerase:

I. synthesizes RNA in the 5' to 3' direction
II. synthesizes RNA in the 3' to 5' direction
III. copies a DNA template
IV. copies an RNA template

A. I and III only
B. II and III only
C. I and IV only
D. II and IV only
E. I only

17. Within primary eukaryotic transcripts, introns are:

A. often functioning as exons in other genes
B. different in size and number among different genes
C. joined to form mature mRNA
D. highly conserved in nucleotide sequence
E. absent from the primary eukaryotic transcript because they are not transcribed

18. Which statement(s) is/are TRUE for eukaryotic protein synthesis?

I. exons of mRNA are spliced together before translation
II. proteins must be spliced soon after translation
III. prokaryotic ribosomes are smaller than eukaryotic ribosomes

A. I only
B. II only
C. I and II only
D. I, II and III only
E. I and III only

19. Which would decrease the transcription of retrotransposons?

I. Acetylation of histones associated with the retrotransposon
II. Deacetylation of histones associated with the retrotransposon
III. Methylation of retrotransposon DNA
IV. Loss of methylation of retrotransposon DNA

A. I and II only
B. I and III only
C. II and III only
D. III and IV only
E. II and IV only

20. Which is an exception to the principle of the *central dogma*?

A. yeast
B. retroviruses
C. bread mold
D. skin cells
E. onion cells

21. miRNA is generated from the cleavage of:

A. double-stranded RNA
B. single-stranded RNA
C. double-stranded DNA
D. single-stranded DNA
E. None of the above

22. The DNA polymerase cannot fully replicate the 3' DNA end, which results in shorter DNA with every division cycle. The new strand synthesis mechanism that follows prevents the loss of the DNA coding region.

Which structure is present at the location of the new strand synthesis?

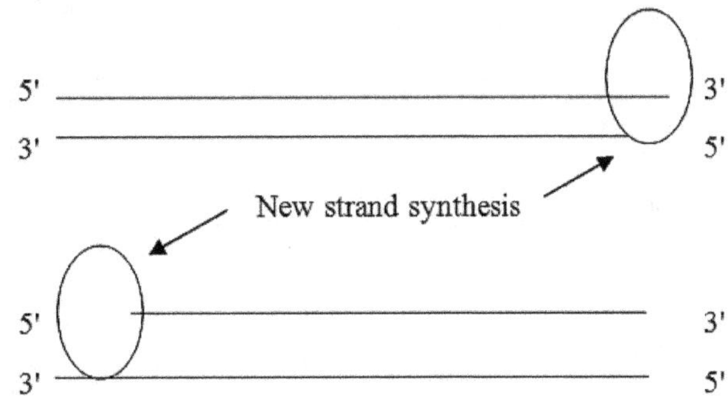

A. kinetochore
B. centrosome
C. telomere
D. centromere
E. chromatid

23. The cell structure composed of a core particle of 8 histones is:

A. telomere
B. nucleosome
C. kinetochore
D. centrosome
E. spindle

24. Which statement is CORRECT about miRNA? (interference RNA):

I. It is a backup system for tRNA in the regulation of translation
II. It base pairs with mRNA and causes it to be cleaved
III. It base pairs with mRNA and prevents its translation

A. II only
B. I and II only
C. II and III only
D. I, II and III
E. I and III only

25. The region of DNA in prokaryotes to which RNA polymerase binds most tightly is the:

A. promoter
B. poly C center
C. enhancer
D. operator site
E. minor groove

26. Genomic libraries contain:

I. promoters
II. intron portions of genes
III. exon portions of genes
IV. retrotransposons

A. I and II only
B. I and III only
C. I, II and III only
D. II and III only
E. I, II, III and IV

27. Which RNA molecules have a common secondary structure called cloverleaf and are relatively small?

A. hnRNA
B. mRNA
C. rRNA
D. tRNA
E. miRNA

28. A chromosome when the linear sequence of a group of genes is reverse of typical sequences has undergone:

A. translocation
B. inversion
C. duplication
D. deletion
E. position effect variegation

29. If a drug inhibits ribosomal RNA synthesis, which eukaryotic organelle is most affected by this drug?

A. Golgi apparatus
B. lysosome
C. mitochondria
D. nucleus
E. nucleolus

30. What evidence shows that the AG gene is vital for forming reproductive organs in *Arabidopsis* flowers?

A. The AG gene encodes a miRNA
B. RNA blot experiments show that the AG gene is strongly expressed in flowers, leaves, and roots
C. The AG gene is in all flowering Arabidopsis plants
D. AG mutant flowers do not have reproductive organs
E. None of the above

31. Which statement about the glycocalyx is FALSE?

A. May be composed of polysaccharide
B. May be composed of polypeptide
C. Protects from osmotic lysis
D. Is used to adhere to surfaces
E. May be responsible for virulence

32. Which does NOT affect chromatin structure?

A. tandem repeats
B. DNA acetylation
C. histone acetylation
D. chromatin remodeling proteins
E. DNA methylation

33. Which enzyme maintains and regulates normal DNA coiling?

A. ligase
B. helicase
C. DNA polymerase I
D. DNA polymerase III
E. topoisomerase

34. RNA polymerase uses the two ribonucleotide triphosphates shown below to make 5'-CG-3'. Which of the indicated phosphorous atoms participates in phosphodiester bond formation?

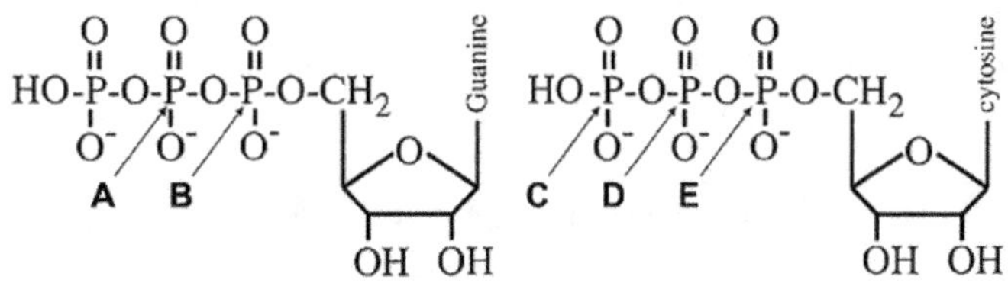

A. phosphorous atom A
B. phosphorous atom B
C. phosphorous atom C
D. phosphorous atom D
E. phosphorous atom E

35. When eukaryotic mRNA hybridizes with its corresponding DNA coding strand (i.e., heteroduplex analysis) and is visualized by electron microscopy, the looping strands of nucleic acid which are seen represent:

A. introns
B. exons
C. lariat structures
D. inverted repeats
E. overlapping genes

36. Which statement does NOT accurately describe an aspect of the nucleosome?

A. has an octet of proteins
B. has a histone H1
C. is the first step in compacting the DNA in the nucleus
D. has DNA wrapped on the outside
E. has a histone H3

37. Which is NOT a bacterial cell wall chemical component?

A. N-acetylmuramic acid
B. peptidoglycan
C. teichoic acids
D. peptide chains
E. cellulose

38. Combinatorial control of gene transcription in eukaryotes is when:

I. each transcription factor regulates only one gene
II. a single transcription factor regulates a combination of genes
III. presence or absence of a combination of transcription factors is required

A. I only
B. II only
C. II and III only
D. III only
E. I, II and III

39. Which post-transcriptional modifications have matured eukaryotic mRNAs undergone before being transported into the cytoplasm?

A. addition of 3' G-cap and 5' poly-A-tail, removal of introns, and splicing of exons
B. RNA splicing together of exons and removal of introns
C. RNA addition of 5' cap and 3' poly-A-tail
D. addition of 5' G-cap and 3' poly-A-tail, removal of introns, and splicing of exons
E. RNA splicing together introns and removal of exons

40. Which statement about gene expression is correct?

A. The ribosome binding site lies at the 3' end of the mRNA
B. The second round of transcription can begin before the preceding transcript is completed
C. Only one gene can be present within a given DNA sequence
D. Mistakes in transcription are corrected by RNA polymerase
E. Change in genotype always results in a changed phenotype

41. Considering that in vitro, the transcription factor SP1 binds nucleic acids with high affinity, where would the radio-labeled SP1 likely NOT be found?

A. Golgi apparatus
B. mitochondria
C. nucleolus
D. ribosomes
E. nucleus

42. During splicing, the snRNA base pairs with:

A. mRNA sequences in the intron
B. hnRNA sequences in the exon
C. DNA sequences in the intron
D. DNA sequences in the exon
E. hnRNA sequences in the intron

43. Which macromolecule would be repaired rather than degraded?

A. triglyceride
B. polynucleotide
C. polypeptide
D. polysaccharide
E. proteins

44. The poly-A tail of RNA is:

A. encoded in the DNA sequence of the gene
B. added by the ribosome during translation
C. base paired with tRNA during translation initiation
D. enzymatically added soon after transcription is finished
E. in the cytoplasm of the cell

45. Which type of histone is not part of the nucleosome core particle?

- **A.** H3
- **B.** H2B
- **C.** H2A
- **D.** H4
- **E.** H1

46. During splicing, the phosphodiester bond at the upstream exon/intron boundary is hydrolyzed by:

- **A.** protein within the snRNP complex
- **B.** 2'-OH of a base within the intron
- **C.** 3'-OH of a base within the intron
- **D.** RNA polymerase III
- **E.** 3'-OH of a base within the exon

47. What is the expected charge, if any, on a histone that binds to DNA?

- **A.** neutral
- **B.** depends on the DNA conformation
- **C.** positive
- **D.** negative
- **E.** neutral or negative

48. RNA polymerase uses the two ribonucleotide triphosphates shown below to make 5'-CG-3'. Which of the indicated oxygen atoms participates in phosphodiester bond formation?

- **A.** oxygen atom A
- **B.** oxygen atom B
- **C.** oxygen atom C
- **D.** oxygen atom D
- **E.** either oxygen atom B or D

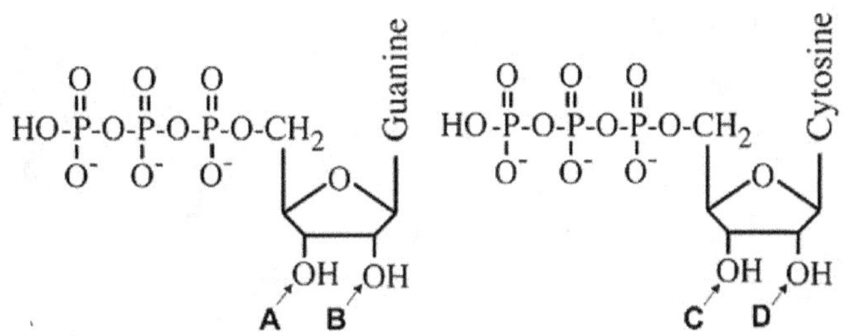

49. Chromosome regions with very few functional genes are:

- **A.** heterochromatin
- **B.** mid-repetitive sequences
- **C.** euchromatin
- **D.** chromatids
- **E.** nucleosomes

50. Which is characteristic of prokaryotes only?

- **A.** primary transcripts of RNA have introns
- **B.** the processed RNA has a poly-A
- **C.** the processed RNA has a 5'-cap
- **D.** transcription of the RNA occurs simultaneously with translation for the RNA
- **E.** the primary transcript is longer than the mRNA

51. What is the product of combining DNA from different sources?

A. recombinant DNA
B. clone
C. hybrid DNA
D. mutant
E. plasmid

52. Transformation of a plant cell is successful when:

A. plasmid cannot enter the cell
B. cell produces daughter cells that subsequently produce other daughter cells
C. cell goes through programmed cell death (apoptosis)
D. cell's enzymes destroy the plasmid after it enters the cell
E. foreign DNA is integrated into one of the plant cell's chromosomes

53. The figure shown illustrates:

A. PCR making a copy of the DNA
B. enzyme cutting the DNA
C. use of hybridization in genetic engineering
D. DNA sequencing via gel electrophoresis
E. insertion of a genetic marker

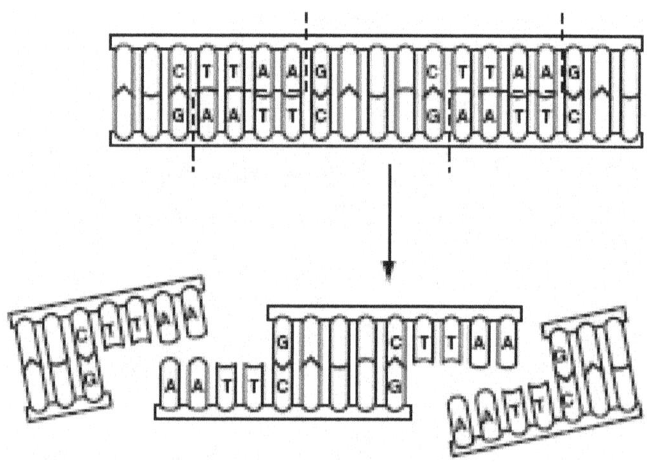

54. Between which two nucleotides are DNA cut in the figure?

A. thymine and adenine
B. adenine and guanine
C. adenine and cytosine
D. thymine and guanine
E. thymine and cytosine

Notes for active learning

Notes for active learning

Biomolecules

1. The linear sequence of amino acids along a peptide chain determines its:

A. primary structure
B. secondary structure
C. tertiary structure
D. quaternary structure
E. none of the above

2. There are [ ] different types of major biomolecules used by humans.

A. a few dozen
B. four
C. several thousand
D. several million
E. a few hundred

3. Amino acids are linked to one another in a protein by which of the following bonds?

A. amide bonds
B. carboxylate bonds
C. ester bonds
D. amine bonds
E. glycosidic bonds

4. What type of protein structure corresponds to a spiral alpha-helix of amino acids?

A. primary
B. secondary
C. tertiary
D. quaternary
E. none of the above

5. Proteins are characterized by the fact that they:

A. always have quaternary structures
B. retain their conformation above 35-40 °C
C. have a primary structure formed by covalent linkages
D. are composed of a single peptide chain
E. none of the above

6. Members of which class of biomolecules is the building blocks of proteins?

A. fatty acids
B. amino acids
C. glycerols
D. monosaccharides
E. nucleic acids

7. Proteins are polymers consisting of which monomer units?

A. keto acids
B. amide
C. amino acids
D. ketones
E. polyaldehydes

8. Why might a change in pH cause a protein to denature?

A. The hydrogen bonds between the hydrophobic portions of the protein collapse due to extra protons
B. The disulfide bridges open
C. The functional groups that give the protein its shape becomes protonated or deprotonated
D. The water hardens and causes the protein's shape to change
E. All the above

9. Collagen is an example of a(n):

A. storage protein
B. transport protein
C. enzyme
D. structural protein
E. hormone

10. The coiling of a chain of amino acids describes a protein's:

A. primary structure
B. secondary structure
C. tertiary structure
D. quaternary structure
E. none of the above

11. What is the name given to the localized bending and folding of a polypeptide backbone of a protein molecule?

A. primary structure
B. secondary structure
C. tertiary structure
D. quaternary structure
E. zymogen structure

12. Which of the following macromolecules are composed of polypeptides?

A. amino acids
B. proteins
C. carbohydrates
D. fats
E. steroids

13. What is the purpose of the plasma membrane?

A. Storing of the genetic material of the cell
B. Retaining water in the cell to prevent it from dehydrating
C. Acting as a cell wall to give the cell structure and support
D. Acting as a boundary, but letting molecules in and out
E. None of the above

14. It is important to have cholesterol in one's body because:

A. it breaks down extra fat lipids
B. it serves as the starting material for the biosynthesis of most other steroids
C. it is the starting material for the building of glycogen
D. the brain is made almost entirely of cholesterol
E. it mainly serves as an energy reserve

15. Cholesterol belongs to the [ ] group of lipids.

A. prostaglandin
B. triacylglycerol
C. saccharides
D. steroid
E. wax

16. The function of cholesterol in a cell membrane is to:

A. act as a precursor to steroid hormones
B. take part in the reactions that produce bile acids
C. maintain structure due to its flat rigid characteristics
D. attract hydrophobic molecules to form solid deposits
E. none of the above

17. The biochemical roles of lipids are:

A. short-term energy storage, transport of molecules, and structural support
B. storage of excess energy, component of cell membranes, and chemical messengers
C. catalysis, protection against outside invaders, motion
D. component of cell membranes, catalysis, and structural support
E. neurotransmitters, hormones, transport of molecules

18. Which of the following molecules is an omega-3 fatty acid?

A. oleic acid
B. linolenic acid
C. linoleic acid
D. palmitic acid
E. none of the above

19. Which of the following is a lipid?

A. lactose
B. aniline
C. nicotine
D. estradiol
E. collagen

20. Lipids are compounds soluble in:

A. glucose solution
B. organic solvents
C. distilled water
D. normal saline solution
E. oxygen

21. Which statement regarding fatty acids is NOT correct? Fatty acids:

A. are always liquids
B. are long-chain carboxylic acids
C. are usually unbranched chains
D. usually have an even number of carbon atoms
E. none of the above

22. Unsaturated triacylglycerols are usually [ ] because [ ]?

A. liquids … they have relatively short fatty acid chains
B. liquids … the kinks in their fatty acid chains prevent their fitting closely
C. liquids … they contain impurities from their natural sources
D. solids … they have relatively long fatty acid chains
E. solids … the similar zig-zag shape of their fatty acid chains allows them to fit closely

23. Oils are generally [ ] at room temperature and are obtained from [ ]:

A. liquids … plants
B. liquids … animals
C. solids … plants
D. solids … animals
E. none of the above

24. Lipids are naturally occurring compounds which all:

A. contain fatty acids as structural units
B. are water-insoluble but soluble in nonpolar solvents
C. contain ester groups
D. contains cholesterol
E. are unsaturated

25. Saturated fats are [ ] at room temperature and are obtained from [ ]?

A. liquids; plants
B. liquids; animals
C. solids; plants
D. solids; animals
E. solids; plants and animals

26. How many molecules of fatty acid are needed to produce one molecule of fat or oil?

A. 1
B. 1.5
C. 2
D. 3
E. 6

27. How many fatty acids are in a phospholipid molecule?

A. 0
B. 1
C. 2
D. 3
E. variable

28. Which of the following is NOT a function of lipids within the body?

A. cushioning to prevent injury
B. insulation
C. energy reserve
D. precursor for glucose catabolism
E. precursor for the synthesis of androgens

29. Which of the following statements describes most monosaccharides?

A. They are unsaturated compounds
B. They are rarely monomers in nature
C. They are composed of carbon, hydrogen, and oxygen, with each carbon bound to at least one oxygen
D. They are insoluble
E. None of the above

30. The three elements in carbohydrates are [ ], [ ] and [ ]:

A. nitrogen, oxygen, hydrogen
B. carbon, hydrogen, oxygen
C. carbon, hydrogen, water
D. nitrogen, oxygen, carbon
E. carbon, nitrogen, hydrogen

31. Fructose does not break apart into smaller units because it is a(n):

A. monosaccharide
B. polysaccharide
C. hexose
D. aldose
E. disaccharide

32. Which of the following molecules is a disaccharide?

A. fructose
B. cellulose
C. amylose
D. glucose
E. lactose

33. What is the major biological function of the glycogen biomolecule?

A. It is used to synthesize disaccharides
B. It is the building block of proteins
C. It stores glucose in animal cells
D. It is a storage form of sucrose
E. It stores glucose in plant cells

34. A carbohydrate that gives two molecules when it is completely hydrolyzed is a:

A. polysaccharide
B. starch
C. monosaccharide
D. disaccharide
E. trisaccharide

35. Which group of carbohydrates CANNOT be hydrolyzed to give smaller molecules?

A. oligosaccharides
B. trisaccharides
C. disaccharides
D. monosaccharides
E. polysaccharides

36. Carbohydrate can be defined as a molecule:

A. composed of carbon atoms bonded to water molecules
B. composed of amine groups and carboxylic acid groups bonded to a carbon skeleton
C. composed mostly of hydrocarbons and soluble in non-polar solvents
D. that is an aldehyde or ketone and has more than one hydroxyl group
E. ending in ~*ase*

37. Disaccharides are best characterized as:

A. two monosaccharides linked by a nitrogen bond
B. two peptides linked by a hydrogen bond
C. two monosaccharides linked by an oxygen bond
D. two amino acids linked by a peptide bond
E. two glycogens linked by a fatty acid

38. The two strands of DNA in the double helix are held by:

A. dipole–dipole attractions
B. metallic bonds
C. ionic bonds
D. covalent bonds
E. none of the above

39. What happens to DNA when placed into an aqueous solution at physiological pH?

A. Individual DNA molecules repel each other due to the presence of positive charges
B. DNA molecules bind to negatively charged proteins
C. Individual DNA molecules attract each other due to the presence of positive and negative charges
D. Individual DNA molecules repel each other due to the presence of negative charges
E. DNA molecules bind to neutral proteins

40. The nucleotide sequence, T-A-G, stands for

A. threonine-alanine-glutamine
B. thymine-adenine-guanine
C. tyrosine-asparagine-glutamic acid
D. thymine-adenine-glutamine
E. none of these

41. If one strand of a DNA double helix has the sequence AGTACTG, what is the sequence of the other strand?

A. GACGTCA
B. AGTACTG
C. GTCATGA
D. TCATGAC
E. AGUACUG

42. The main role of DNA is to provide instructions on how to build:

I. lipids II. carbohydrates III. proteins

A. I only
B. II only
C. III only
D. I and II only
E. I, II and III

43. Nucleic acids are polymers of [ ] monomers.

A. monosaccharide
B. fatty acid
C. DNA
D. nucleotide
E. none of the above

44. During DNA transcription, a guanine base on the template strand codes for which base on the growing RNA strand?

A. guanine
B. thymine
C. adenine
D. cytosine
E. uracil

45. What type of biological compound is a polymer composed of sugar, a base, and phosphoric acid?

A. nucleic acid
B. lipid
C. carbohydrate
D. protein
E. none of the above

46. The number of adenines in a DNA molecule is equal to the number of thymines because:

A. adenines are paired opposite of guanine in a DNA molecule
B. of the strong attraction between the nucleotides of adenine and thymine
C. the structure of adenine is similar to uracil
D. adenine is paired to cytosine in a DNA molecule
E. none of the above

47. What is the sugar component in RNA called?

A. fructose
B. galactose
C. glucose
D. ribose
E. deoxyribose

48. During DNA replication, an adenine base on the template strand codes for which base on the complementary strand?

A. thymine
B. guanine
C. cytosine
D. adenine
E. uracil

49. Which of the following is NOT part of a nucleotide?

A. cyclic nitrogenous base
B. fatty acid
C. phosphate group
D. cyclic sugar
E. oxygen

50. Which of the following linkage is in a nucleic acid?

A. phosphate linkage
B. ester linkage
C. glycoside linkage
D. peptide linkage
E. amide linkage

51. The one cyclic amine base that occurs in DNA but not in RNA is:

A. cystine
B. guanine
C. thymine
D. uracil
E. adenine

52. In the synthesis of mRNA, an adenine in the DNA pairs with:

A. guanine
B. thymine
C. uracil
D. adenine
E. cytosine

53. Nucleic acids determine the:

A. quantity and type of prions
B. number of mitochondria in a cell
C. sequence of amino acids
D. pH of the cell nucleus
E. catabolism rate for food

Notes for active learning

Notes for active learning

Specialized Cells and Tissues

1. What is the function of Ca^{2+} in muscle contractions?

   A. Breaking the cross-bridges as a cofactor in the hydrolysis of ATP
   B. Re-establishing the polarity of the plasma membrane following an action potential
   C. Transmitting the action potential across the neuromuscular junction
   D. Spreading the action potential through the T tubules
   E. Binding to the troponin complex, which leads to exposure of myosin-binding sites

2. As sodium moves from the extracellular to intracellular space, which anion follows for electrical neutrality?

   A. chloride
   B. potassium
   C. magnesium
   D. lithium
   E. calcium

3. Which statement about caffeine is FALSE?

   A. It inhibits the signaling pathway normally stimulated by epinephrine
   B. It is a signal molecule
   C. It acts in different ways in different tissues
   D. It is a common ingredient in headache remedies
   E. It indirectly leads to an increased rate of conversion of glycogen into glucose

Use the graph to answer questions **4–8**

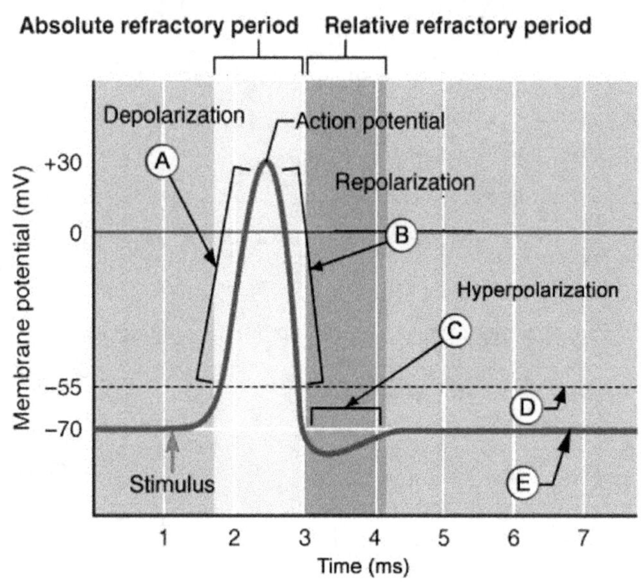

4. In the figure provided, arrow A points to which ion enters the axon?

   A. Ca^{2+}
   B. K^+
   C. Cl^-
   D. Na^+
   E. Mg^{2+}

5. In the figure provided, arrow B points to which ion leaving the axon?

   A. Ca^{2+}
   B. K^+
   C. Cl^-
   D. Na^+
   E. Mg^{2+}

6. From the figure provided, arrow C points to the slow close of:

A. ATPase
B. Mg^{2+} voltage-gated channels
C. Cl^- voltage-gated channels
D. Na^+ voltage-gated channels
E. K^+ voltage-gated channels

7. From the figure provided, arrow D points to:

A. hyperpolarization
B. threshold
C. overshoot
D. resting potential
E. summation

8. From the figure provided, arrow E points to:

A. hyperpolarization
B. hypopolarization
C. overshoot
D. resting potential
E. summation

9. What is the effect of chemical X that denatures all enzymes in the synaptic cleft?

A. inhibition of depolarization of the presynaptic membrane
B. prolonged depolarization of the presynaptic membrane
C. prolonged depolarization of the postsynaptic membrane
D. inhibition of depolarization of the postsynaptic membrane
E. inhibition of communication between the presynaptic and postsynaptic neurons

10. A "resting" motor neuron is expected to:

A. release elevated levels of acetylcholine
B. have high permeability to Na^+
C. have equal permeability to Na^+ and K^+
D. exhibit a resting potential that is more negative than the threshold potential
E. have a higher concentration of Na^+ on the inside of the cell than on the outside

11. Which results from the administration of digitalis known to block Na^+/K^+ ATPase?

A. decrease of intracellular $[K^+]$
B. increase of extracellular $[Na^+]$
C. decrease of intracellular $[Ca^{2+}]$
D. decrease of intracellular $[Na^+]$
E. increase of intracellular $[K^+]$

12. If a neuron's membrane potential goes from –70 mV to –90 mV, this is an example of:

A. depolarization
B. repolarization
C. hyperpolarization
D. Na^+ channel inactivation
E. hyperpolarization

13. Which tissue is an example of connective tissue?

I. bone II. cartilage III. nervous

A. II only
B. I only
C. II and III only
D. I, II and III
E. I and II only

14. Which factor does NOT affect the resting membrane potential?

A. The Na^+/K^+ pump
B. Active transport across the plasma membrane of the axon
C. Equal distribution of ions across the plasma membrane of the axon
D. Selective permeability for ions across the plasma membrane of the axon
E. The relative concentration of ions across the plasma membrane

15. The first event to occur when a resting axon reaches its threshold potential is:

A. closing of K^+ gates
B. activation of the Na^+/K^+ pump
C. hyperpolarization of the membrane potential
D. closing of Na^+ gates
E. opening of Na^+ gates

16. Multiple sclerosis is a demyelinating disease most severely affecting which muscle fibers?

A. C-pain – temperature and mechanoreception (velocity 0.75-3.5 m/s)
B. gamma – intrafusal muscle spindle (velocity 4-24 m/s)
C. beta – touch (velocity 25-75 m/s)
D. alpha – extrafusal muscle spindle (velocity 75-130 m/s)
E. beta – pressure (velocity 25-75 m/s)

17. The measure of the postsynaptic response after stimulation by the presynaptic cell is called synaptic:

A. area
B. strength
C. speed
D. summation
E. refractory

18. What molecule must a cell have to be able to respond to a signal?

A. paracrine
B. receptor
C. autocrine
D. responder
E. A and C

19. Muscle contraction occurs due to:

A. myosin filaments and actin filaments sliding past each other
B. shortening of only actin filaments
C. shortening of only myosin filaments
D. simultaneous shortening of both myosin and actin filaments
E. myosin filaments elongate, while actin filaments shorten

20. When an organism dies, its muscles remain in the contracted state of *rigor mortis* for a brief period. Which most directly contributes to this phenomenon?

A. no ATP is available to move cross-bridges
B. no ATP is available to break bonds between the thick and thin filaments
C. no calcium to bind to troponin
D. no oxygen supplied to muscles
E. glycogen remaining in the muscles

21. Which would be expected in a cell exposed to ouabain, known to block the activity of the Na^+/K^+ ATPase?

A. increase in ATP consumption
B. spontaneous depolarization
C. increase in extracellular $[Na^+]$
D. increase in intracellular $[K^+]$
E. increase in extracellular $[Na^+]$ and intracellular $[K^+]$

22. The region on neurons that brings the graded potential to the neuronal cell body is the:

A. axon
B. dendrite
C. T-tubule
D. node of Ranvier
E. axon terminus

23. Which statement correctly describes the function of the myelin sheath in action potential transmission?

A. Saltatory conduction dissipates current through specialized leakage channels
B. Oligodendrocytes cover the nodes of Ranvier to prevent backflow of current
C. Protein fibers cover the axon and prevent leakage of current across the membrane
D. Nodes of Ranvier insulate the axon
E. Lipids insulate the axons, while membrane depolarization occurs within the nodes

24. Which human structure is analogous to an electrical device allowing current to flow only in one direction?

A. dendrite
B. axon process
C. myelin sheath
D. synaptic cleft
E. spinal nerve

25. In the communication link between a motor neuron and a skeletal muscle:

A. motor neuron is presynaptic, and the skeletal muscle is the postsynaptic cell
B. motor neuron is postsynaptic, and the skeletal muscle is the presynaptic cell
C. action potentials are possible on the motor neuron but not on the skeletal muscle
D. action potentials are possible on the skeletal muscle but not on the motor neuron
E. motor neuron fires action potentials, but the skeletal muscle is not electrochemically excitable

26. Which organelle undergoes self-replication?

A. DNA
B. mitochondria
C. nucleolus
D. ribosomes
E. nucleus

27. The resting membrane potential of a neuron would be closest to:

A. 70 mV
B. 120 mV
C. –70 mV
D. 0 mV
E. 30mV

28. A molecule that binds to the three-dimensional structure of another molecule's receptor is:

A. responder
B. receptor
C. ligand
D. ion channel
E. filament

29. *Myasthenia gravis* is a severe autoimmune disease of neuromuscular junctions. A patient produces antibodies against the acetylcholine receptors of the muscle membrane (i.e., the sarcolemma), causing their removal by phagocytosis. Which process would be directly affected by this condition?

A. Acetylcholine synthesis
B. Calcium release by the endoplasmic reticulum
C. Sarcomere shortening during muscle contraction
D. Density of the receptors on the presynaptic membrane
E. Action potential conduction across the sarcolemma

30. Which are shared by skeletal, cardiac, and smooth muscles?

A. A bands and I bands
B. transverse tubules
C. gap junctions
D. thick and thin filaments
E. motor units

31. Procaine is a local anesthetic used during many dental procedures that inhibit the propagation of an action potential along a neuron by:

A. increasing the myelin deposit of Schwann cells
B. removing Schwann cells covering the axon
C. increasing Cl^- movement out of the neuron in response to an action potential
D. stimulating Ca^{2+} voltage-gated channels at the synapse
E. blocking Na^+ voltage-gated channels

32. The glial cell that myelinates an axon in the CNS is the:

A. astrocyte
B. oligodendrocyte
C. Schwann Cell
D. choroid plexus
E. chondrocyte

33. The myelin sheath around axons of the peripheral nervous system is produced by:

A. axon Hillock
B. nerve cell body
C. nodes of Ranvier
D. Schwann cell
E. oligodendrocytes

34. All these statements are true for muscles, EXCEPT:

A. Tetanus is a condition of sustained contraction due to an overlap of twitch impulses
B. Tonus is the state of partial contraction which occurs in a resting muscle
C. Isometric contraction means that the length of the muscle is constant
D. Isotonic contraction means that the length of the muscle shortens
E. Resting muscles are completely relaxed

35. Tissues are composed of cells, and a group of tissues functioning together makes up:

A. organs
B. membranes
C. organ systems
D. organelles
E. organisms

36. If a neuronal membrane, which is usually slightly passively permeable to K^+, becomes impermeable, but the Na^+/K^+-ATPase remains active, the neuron's resting potential would become:

A. less positive because $[K^+]$ increases inside the neuron
B. more negative because $[K^+]$ increases outside the neuron
C. more positive because $[K^+]$ increases inside the neuron
D. more positive because $[K^+]$ increases outside the neuron
E. more negative because $[K^+]$ increases inside the neuron

37. The asymmetric concentration gradient of Na^+ and K^+ across the membrane is maintained by the:

A. voltage-gated channels
B. Na^+/K^+ ATPase
C. mitochondria
D. constitutive ion channels
E. Schwann cells

38. Which statement about the insulin receptor is FALSE?

A. Once activated, it undergoes autophosphorylation
B. It requires binding by two insulin molecules to be activated
C. It catalyzes the phosphorylation of the insulin response substrate
D. It is entirely within the cytoplasm
E. None of the above

39. Which process results in hyperpolarization?

A. excessive outflow of K^+
B. excessive outflow of Na^+
C. excessive influx of K^+
D. excessive influx of Na^+
E. excessive outflow of Na^+ and an excessive influx of K^+

40. The deficiency of which vitamin is associated with neural tube defects?

A. B_6
B. calcium
C. folic acid
D. B_{12}
E. B_7

41. Which membrane-bound protein channel must be inhibited to cause a blockage of nerve conduction by a hydrophobic local anesthetic such as lidocaine?

A. Cl^- channel
B. Ca^{2+} channel
C. K^+ channel
D. Na^+ channel
E. Mg^{2+} channel

42. Which is NOT a class of molecules used as neurotransmitters?

A. peptides
B. catecholamines
C. amino acids
D. neuropeptides
E. enzymes

43. Which is FALSE?

A. Cells are bombarded with numerous signals, but they respond to only a few
B. A cell's receptors determine whether the cell will respond to a signal
C. Receptor proteins are very specific
D. There are only a few kinds of signal receptor proteins
E. None of the above

44. Which neurons are involved in the piloerection of hair standing on its end?

A. sympathetic motor neurons
B. sympathetic sensory neurons
C. parasympathetic motor neurons
D. parasympathetic sensory neurons
E. sympathetic and parasympathetic sensory neurons

45. Which statement best summarizes the relationship between the cytoplasm and cytosol?

A. Cytoplasm includes the cytosol, a watery fluid inside the cell
B. Cytoplasm is within the nucleolus, while cytosol is within the cell outside the nucleus
C. Cytoplasm is within the cell outside the nucleus, while cytosol is within the nucleolus
D. Cytoplasm is within the nucleus, while cytosol is within the nucleolus
E. Cytoplasm is within the nucleolus, while cytosol is within the mitochondria

46. During depolarization of a muscle cell, Ca^{2+} is released from the sarcoplasmic reticulum and binds to:

A. muscle ATPase
B. troponin C
C. myosin heads
D. actin thin filaments
E. myosin thick filaments

47. Which is TRUE about saltatory conduction?

A. Current passes through the myelin sheath
B. Voltage-gated Ca^{2+} channels are concentrated at the nodes of Ranvier
C. Myelinated axons exhibit greater conduction velocity than non-myelinated axons
D. It is much slower than conduction along non-myelinated axons
E. The direction of depolarization is reversed

48. The specialized region of the neuron that connects to the axon and sums graded inputs prior to propagation of *all or none* depolarization is:

A. axon hillock
B. nerve cell body
C. node of Ranvier
D. glial cell
E. axon terminus

49. Myelin covers axons and is responsible for the following:

A. initiating the action potential
B. allowing pumping of Na^+ out of the cell
C. maintaining the resting potential
D. determining the threshold of the neuron
E. allowing faster conduction of impulses

50. Which tissue forms coverings, linings, and glands?

A. connective tissue
B. cellular matrix
C. endothelium
D. epithelial
E. peritoneum

51. What is the function of the nodes of Ranvier within the neuron?

A. To provide a binding site for acetylcholine
B. To provide a space for Schwann cells to deposit myelin
C. To regenerate the anterograde conduction of the action potential
D. To permit the axon hillock to generate a stronger action potential
E. To permit the axon process to generate an action potential with a greater amplitude

52. An organ is a structure with a recognizable shape and specific functions and is composed of different:

A. tissues
B. cells
C. germ layers
D. mesoderm
E. endoderm

53. What can be deduced about the conduction velocity of Purkinje fibers?

A. Fast, ion independent channel
B. Slow, ion independent channel
C. Fast, Na^+ dependent channel
D. Slow, Na^+ dependent channel
E. none of the above

54. Which is NOT classified as one of the four primary (basic) tissue types?

A. connective
B. blood
C. muscle
D. nervous
E. epithelial

55. The release of a neurotransmitter into the synaptic cleft results from the influx of:

A. Na^+ ions
B. K^+ ions
C. Mg^{2+} ions
D. Cl^- ions
E. Ca^{2+} ions

56. Which connective tissue stores triglycerides and provides cushioning and support for organs?

A. glycogen
B. muscle
C. connective
D. adipose
E. endothelial

57. Which cells do NOT utilize voltage-gated Na^+ channels?

A. cardiac cells
B. endothelial cells
C. smooth muscle cells
D. skeletal muscle cells
E. nerve cells

58. The cells lining the air sacs in the lungs are:

A. simple columnar epithelium
B. highly elastic connective tissue
C. stratified squamous epithelium
D. pseudostratified ciliated columnar epithelium
E. simple squamous epithelium

Notes for active learning

Notes for active learning

Microbiology

1. What is the *major* distinction between prokaryotic and eukaryotic cells?

   A. prokaryotic cells do not have DNA, and eukaryotic cells do
   B. prokaryotic cells cannot obtain energy from their environment
   C. eukaryotic cells are smaller than prokaryotic cells
   D. prokaryotic cells have not prospered, while eukaryotic cells are evolutionary "successes."
   E. prokaryotic cells do not have a nucleolus, but eukaryotic cells do

2. Which statement describes the actions of penicillin?

   A. it is a reversible competitive inhibitor
   B. it is an irreversible competitive inhibitor
   C. it activates transpeptidase that digests the bacterial cell wall
   D. it is an effective antiviral agent
   E. it acts as a noncompetitive inhibitor

3. Which statement applies to all viruses?

   A. They have an RNA genome
   B. They have a DNA genome
   C. They have chromosomes
   D. They cannot replicate outside of a host cell
   E. They have reverse transcriptase

4. The replica plating technique of Joshua and Esther Lederberg demonstrated that:

   A. mutations are usually beneficial
   B. mutations are usually deleterious
   C. streptomycin caused the formation of streptomycin-resistant bacteria
   D. streptomycin revealed the presence of streptomycin-resistant bacteria
   E. the frequency of mutations is proportional to the concentration of streptomycin

5. Operons:

   A. are a common feature of the eukaryote genome
   B. often coordinate the production of enzymes that function in a single pathway
   C. have multiple translation start and stop sites used by ribosomes
   D. usually undergo alternative splicing
   E. both B and C

6. Which is TRUE for the life cycle of sexually reproducing *Neurospora* fungus?

A. Only mitosis occurs
B. Fertilization and meiosis are separated
C. Meiosis quickly follows fertilization
D. Fertilization immediately follows meiosis
E. Mitosis quickly follows fertilization

7. Which is in prokaryotic cells?

A. mitochondria
B. chloroplasts
C. nuclei
D. enzymes
E. extensive endomembrane system

8. Which organelle is the site of protein modification and carbohydrate synthesis?

A. Golgi apparatus
B. lysosomes
C. peroxisomes
D. smooth ER
E. nucleolus

9. What is likely to occur if a suspension of Hfr cells is mixed with excess F^- cells?

A. Most of the F^- cells are transformed into F^+ cells
B. The F^- cells produce sex pili that attach to the Hfr cells
C. Hfr chromosomal DNA is transferred to F^- cells by conjugation
D. Hfr cells replicate the F factor independently of their chromosomes
E. Most of the F^+ cells become F^- cells

10. What carcinogen and mutagen test looks for an increased reversion frequency in a His^- bacteria strain?

A. *Salmonella* reversion test
B. auxotrophic reversion test
C. mutagen test
D. amber test
E. Ames test

11. The type of bacteria NOT able to grow on minimal media due to mutations affecting metabolism:

A. auxotrophs
B. chemotrophs
C. heterotrophs
D. prototrophs
E. all are able to grow

12. Which statement is TRUE?

A. Endospores are for reproduction
B. Endospores allow a cell to survive environmental changes
C. Endospores are easily stained with Gram stain
D. Cell produces one endospore and keeps growing
E. Cell produces many endospores and keeps growing

13. Most fungi spend the biggest portion of their life cycle as:

A. neither haploid nor diploid
B. both haploid and diploid
C. diploid
D. polyploidy
E. haploid

14. An aerobic bacteria culture that has been exposed to cyanide gas is infected by a bacteriophage strain. However, replication of viruses does not occur. What is cyanide's action mechanism?

A. Binding to viral nucleic acid
B. Denaturing bacteriophage enzymes
C. Inhibiting aerobic ATP production
D. Destroying bacteriophage binding sites on the bacterial cell wall
E. Denaturing viral enzymes needed for replication

15. Recipient cells acquire genes from free DNA molecules in the surrounding medium by:

A. generalized transduction
B. conjugation
C. transduction
D. recombination
E. transformation

16. Viruses can have a genome that is:

A. single-stranded DNA
B. single-stranded RNA
C. double-stranded RNA
D. double-stranded DNA
E. all the above

17. Which assumption must be true to map the order of bacterial genes on the chromosome in a Hfr strain?

A. Bacterial genes are polycistronic
B. A given Hfr strain always transfers its genes in the same order
C. The rate of chromosome transfer varies between bacteria of the same strain
D. Different mechanisms replicate the inserted F factor and bacterial genes
E. All the above statements are true

18. Which statement about prokaryotic cells is generally FALSE?

A. They have a semirigid cell wall
B. They are motile using flagella
C. They possess 80S ribosomes
D. They reproduce by binary fission
E. They lack membrane-bound nuclei

19. All events have a role in the life cycle of a typical retrovirus, EXCEPT:

A. injection of viral DNA into the host cell
B. integration of viral DNA into the host genome
C. reverse transcriptase gene is transcribed, and mRNA is translated inside the host cell
D. viral DNA incorporated into the host genome may be replicated along with the host DNA
E. none of the above

20. All are true for viruses EXCEPT:

A. Genetic material may be either single-stranded or double-stranded RNA
B. A virus may replicate in a bacterial or eukaryotic host
C. A virus may replicate without a host
D. The protein coat of the virus does not enter a host bacterial cell
E. Genetic material may be either single-stranded or double-stranded DNA

21. DNA transfer from a bacterial donor cell to a recipient cell by cell-to-cell contact is:

A. conjugation
B. transformation
C. transduction
D. recombination
E. transposons

22. On the overnight agar plates with *E. coli*, replication of a virus is marked by:

A. no visible change
B. bacterial colonies on the agar surface
C. growth of a smooth layer of bacteria across the plate
D. growth of bacteria across the entire plate except for small clear patches
E. absence of any growth on the plate

23. Which statement about gram-negative cell walls is FALSE?

A. They protect the cell in a hypotonic environment
B. They have an extra outer layer of lipoproteins, lipopolysaccharides, and phospholipids
C. They are toxic to humans
D. They have a thinner outer membrane
E. They are sensitive to penicillin

24. All the following may be present in a mature virus outside the host cell, EXCEPT:

A. core proteins
B. both RNA and DNA
C. protein capsid
D. phospholipid bilayer envelope
E. none of the above are necessary

25. All are correct about *lac* operon, EXCEPT:

A. Repressor protein binds to the operator, halting gene expression
B. The promoter is the binding site of RNA polymerase
C. There is not a gene that encodes for a repressor protein
D. Three structural genes code for functional proteins
E. *Lac* operon is in eukaryotes

26. Phage DNA integrated into the chromosome is:

A. lytic phage
B. specialized transducing phage
C. lysogenic phage
D. prophage
E. insertion sequence

27. Many RNA copies of the retrovirus RNA genome are made by:

A. host cell DNA polymerases
B. reverse transcriptase
C. host cell RNA polymerases
D. host cell ribosomes
E. none of the above

28. In the laboratory, *E. coli* are grown at a temperature of 37 °C because:

A. *E. coli* strain is a 37 °C temperature-sensitive mutant
B. *E. coli* reproduces most rapidly at this temperature
C. lower temperatures inhibit conjugation
D. *E. coli* are obligate aerobes
E. *E. coli* obtains energy from the temperature of the growth medium

29. Some bacteria can propel themselves through liquid using:

A. flagellum
B. centriole
C. centrosome
D. peptidoglycan
E. cell wall

30. When most viruses infect eukaryotic cells:

A. their capsid does not enter the host cell
B. they replicate independently of the host cell during a lysogenic infection
C. they can enter the cell via endocytosis
D. they do not need to have host-specific proteins to infect the target cell
E. they replicate as the host cell replicates during a lytic infection

31. Which organelle(s) is/are NOT present in bacteria?

I. peroxisomes II. nucleolus III. ribosomes IV. flagellum

A. I only
B. II only
C. I and II only
D. I, II, and III
E. III and IV only

32. A bacterial cell carrying a prophage is:

A. virulent
B. temperate
C. exconjugant
D. transformant
E. lysogen

33. All are true of prokaryotic translation, EXCEPT:

A. mRNA is not spliced before initiation
B. N-terminal amino acid of nascent polypeptides is formylated
C. mRNA chain being translated may not be fully transcribed before translation begins
D. Hydrogen bonds between amino acids and mRNA codons are necessary for translation
E. Translation and transcription both happen in the same location within the cell

34. Prokaryotes are about how many times smaller in diameter than a typical eukaryote?

A. two
B. ten
C. 100
D. 10,000
E. zero

35. Which structures are in prokaryotes?

I. A cell wall containing peptidoglycan
II. A plasma membrane with cholesterol
III. Ribosomes

A. I only
B. II only
C. I and II only
D. I and III only
E. II and III only

36. Which is characteristic of viruses?

A. membrane-bound organelles
B. genetic material not made of nucleic acids
C. peptidoglycan cell wall
D. phospholipid bilayer membrane
E. protein coat

37. An F-plasmid that can integrate into the bacterial chromosome by homologous recombination is:

A. episome
B. virion
C. endoconjugate
D. lytic
E. virulent

38. The RNA genome of a retrovirus is converted to double-stranded DNA by:

A. host cell DNA polymerases
B. reverse transcriptase
C. host cell RNA polymerases
D. host cell ribosomes
E. none of the above

39. Which statement is true about T4 infection of *E. coli*?

A. T4 mRNA is translated by bacterial ribosomes while still being transcribed from DNA
B. One of the first genes expressed during viral infection is a lysozyme that facilitates cell lysis
C. The final stage for lytic cycles of infection is the viral assembly after the virus leaves the cell
D. T4 buds via endocytosis through the plasma membrane to leave the cell
E. T4 exits the cell through protein pores in the plasma membrane to leave the cell

40. Which statement about a gram-positive cell wall is FALSE?

A. It maintains the shape of cells
B. It is sensitive to lysozyme
C. It protects the cell in a hypertonic environment
D. It contains teichoic acids
E. It is sensitive to penicillin

41. Which statement is correct about the lipopolysaccharide layer outside the peptidoglycan cell wall of a Gram-negative bacterium?

A. It allows the bacterium to attach to solid objects
B. It does not contain a phospholipid membrane
C. It protects the bacterium against antibiotics
D. It absorbs and holds Gram stain
E. It appears deep purple from Gram stain

42. DNase added to a bacterial cell causes hydrolysis of the cell's DNA, preventing protein synthesis and cell death. Regardless, some viruses pretreated with DNase produce new proteins following infection because:

A. viral genome contains multiple copies of their genes
B. viruses are homozygous for necessary genes
C. viral genome contains multiple reading frames
D. icosahedral protein coat of the virus denatures DNase
E. viral genome is comprised of RNA

43. In a mating between Hfr and F^- cells, the F^- recipient:

A. becomes Hfr
B. becomes F′
C. remains F^-
D. becomes F^+
E. cannot establish lysogeny

44. The unique feature of the methionine residue used for prokaryotic initiation of translation is that it is:

A. formylated
B. hydrophilic
C. methylated
D. acetylated
E. hydrophobic

45. Which statement describes a bacterial cell placed in a 35% large polysaccharide solution (e.g., dextran)?

A. NaCl moves into the cell from a higher to a lower concentration
B. The cell undergoes plasmolysis
C. H_2O moves out of the cell
D. H_2O moves into the cell
E. No change result because the solution is isotonic

46. If the Gram stain method is used to stain a Gram-positive bacterium, it appears?

A. deep purple because of a thicker peptidoglycan cell wall
B. deep purple because of a thinner peptidoglycan cell wall
C. red or pink because of a thicker peptidoglycan cell wall
D. red or pink because of a thinner peptidoglycan cell wall
E. red or pink, because of the absence of a peptidoglycan cell wall

47. Which enzyme replicates the F factor in F^+ bacteria prior to conjugation?

A. DNA polymerase
B. reverse transcriptase
C. DNA ligase
D. integrase
E. RNA polymerase

48. In a mating between Hfr and F^- cells, the Hfr donor:

A. becomes F'
B. remains Hfr
C. becomes F^+
D. becomes $F^{-'}$
E. loses part of the chromosome

49. Which statement(s) is/are TRUE regarding retrotransposons?

I. They are never between genes in the human genome
II. They comprise close to half of the human genome
III. They can cause mutations by inserting themselves into genes

A. I and II only
B. II only
C. I and III only
D. I, II and III
E. II and III only

50. Which is TRUE about plasmids?

I. They are small organelles in the bacterial cytoplasm
II. They are transcribed and translated simultaneously
III. Bacterial enzymes replicate them

A. I only
B. II and III only
C. I and III only
D. I, II, and III
E. I and II only

51. Which has a cell wall?

A. protoplasts
B. fungi
C. L forms
D. viruses
E. mycoplasmas

52. Which statement about a prokaryotic cell is correct?

A. It contains a range of different organelles
B. It has a nucleolus within the cytoplasm
C. It contains cell walls composed of chitin
D. It has a nuclear membrane that encloses a nucleus
E. It uses glycolysis to produce ATP

53. Which statement applies to both a bacteriophage and retrovirus?

A. They are capable of infecting human cells
B. They function as immunosuppressive agents
C. They integrate their genetic material into the genome of the host cell
D. They have genes that encode reverse transcriptase
E. They have a genome of RNA

54. Integration of the phage DNA into the bacterial chromosome is facilitated by:

A. DNA polymerase encoded in the host genome
B. topoisomerase encoded in the phage
C. several proteins, some of them encoded in the phage and some in the host genome
D. site-specific recombinase encoded in the host genome
E. site-specific recombinase encoded in the phage

55. After being digested by restriction enzymes, how is DNA ligated into a plasmid in two different directions?

A. Both ends of a DNA fragment produced by a restriction enzyme are identical if rotated 180°
B. The existing DNA strands serve as primers for DNA polymerase
C. DNA ligase enzymes can link any two pieces of DNA together
D. Plasmid DNA is single-stranded, so the ligated strands form double-stranded segments
E. DNA polymerase links any two pieces of DNA together

56. Virulent phage:

I. is capable only of lytic growth
II. can only undergo a process called lysogeny
III. needs several different proteins to be incorporated into the bacterial chromosome

A. I and II only
B. III only
C. II only
D. I only
E. I, II, and III

57. Which statement best describes the promoter in an operon?

A. It activates the repressor-inducer complex to permit transcription
B. It is a molecule that inactivates the repressor and turns on the operon
C. It is the binding site for RNA polymerase
D. It is the binding site for the repressor
E. It is the binding site for DNA polymerase

58. Bacteria transformed by a plasmid can be distinguished from untransformed bacteria by the following:

A. genetic marker
B. trisomy
C. presence of DNA strands
D. absence of cytosine
E. hybridization

59. Which occurs during transformation?

A. bacterial DNA undergoes mutation
B. bacterial DNA is inserted into a plasmid
C. two bacteria exchange genetic material via pilli
D. a prokaryotic cell becomes eukaryotic
E. a bacterial cell takes in foreign DNA material

60. Plasmids are easily inserted into bacteria but not into yeast because yeast is:

A. mutants
B. protists
C. eukaryotes
D. prokaryotes
E. RNA carriers

Notes for active learning

Section III: DNA, Genes & Heredity

Mechanisms of Reproduction

1. How many double-stranded DNA molecules are in a single mouse chromosome after gametes form?

A. 0
B. 1
C. 2
D. 4
E. 8

2. Common red-green color blindness is an X-linked trait. When a woman whose father is colorblind has a son with a non-afflicted man, what is the probability that their son will be colorblind?

A. 0
B. 1/4
C. 1/2
D. 3/4
E. 1

3. At birth, a woman possesses a finite number of ova. In oogenesis, the meiotic division is arrested at which stage until she reaches menarche?

A. ovum
B. oogonium
C. ova
D. secondary oocytes
E. primary oocytes

4. The likely gamete cell to be produced from meiosis in the seminiferous tubules is:

A. diploid 2° spermatocytes
B. haploid 1° spermatocytes
C. haploid spermatids
D. diploid spermatids
E. none of the above

5. A human cell after the first meiotic division is:

A. 2N and 2 chromatids
B. 2N and 4 chromatids
C. 1N and 2 chromatids
D. 1N and 1 chromatid
E. None of the above

6. What distinguishes meiosis from mitosis?

I. Genetic recombination
II. Failure to synthesize DNA between successive cell divisions
III. Separation of homologous chromosomes into distinct cells

A. I only
B. II only
C. I and III only
D. II and III only
E. I, II and III

7. Oocytes within primordial follicles of the ovary are arrested in:

A. interphase
B. prophase II of meiosis
C. prophase I of meiosis
D. prophase of mitosis
E. G1

8. SRY gene encoding for the testis-determining factor is the primary sex-determining gene that resides on:

A. pseudoautosomal region of the Y chromosome
B. short arm of the Y chromosome, but not in the pseudoautosomal region
C. X chromosome
D. pseudoautosomal region of the X chromosome
E. autosomes

9. The number of chromosomes contained in the human primary spermatocyte is:

A. 23
B. 23, X/23, Y
C. 92
D. 184
E. 46

10. Which process does NOT contribute to genetic variation?

A. Random segregation of homologous chromosomes during meiosis
B. Random segregation of chromatids during mitosis
C. Recombination
D. Mutation
E. All the above

11. In human females, secondary oocytes do not complete meiosis II until:

A. menarche
B. menstruation
C. puberty
D. menopause
E. fertilization

12. 47, XXY is a condition known as:

A. Turner syndrome
B. double Y syndrome
C. trisomy X syndrome
D. Klinefelter syndrome
E. fragile X syndrome

13. All cells contain diploid (2N) numbers of chromosomes EXCEPT:

A. primary oocyte
B. spermatogonium
C. spermatid
D. zygote
E. oogonium

14. The probability that all children in a four-children family will be males is:

A. 1/2
B. 1/4
C. 1/8
D. 1/16
E. 1/64

15. Unequal cytoplasm division is characteristic of:

A. binary fission of bacteria
B. mitosis of a kidney cell
C. production of sperm
D. production of an ovum
E. none of the above

16. All are clinical manifestations of Kartagener's syndrome, resulting from defective dynein that causes paralysis of cilia and flagella, EXCEPT:

A. chronic respiratory disorders
B. cessation of ovulation
C. male infertility
D. ectopic pregnancy
E. middle ear infections

17. Translation, transcription, and replication occur in which phase of the cell cycle?

A. G1
B. G2
C. metaphase
D. anaphase
E. S

18. Progesterone is primarily secreted by the:

A. primary oocyte
B. hypothalamus
C. corpus luteum
D. anterior pituitary gland
E. endometrial lining

19. All statements are true for a normal human gamete, EXCEPT it:

A. originates via meiosis from a somatic cell
B. contains genetic material that has undergone recombination
C. contains a haploid number of genes
D. always contains an X or Y chromosome
E. forms from the meiotic process

20. Which cell division results in four genetically different daughter cells with 1N chromosomes?

A. interphase
B. somatic cell regeneration
C. cell division
D. mitosis
E. meiosis

21. A genetically important event of crossing over occurs during:

A. metaphase II
B. telophase I
C. anaphase I
D. prophase I
E. telophase II

22. Which does NOT describe events occurring in prophase I of meiosis?

A. chromosomal migration
B. genetic recombination
C. formation of a chiasma
D. spindle apparatus formation
E. tetrad formation

23. The difference between spermatogenesis and oogenesis is:

A. spermatogenesis produces haploid cells, while oogenesis produces diploid cells
B. spermatogenesis produces gametes, while oogenesis does not produce gametes
C. oogenesis is a mitotic process, while spermatogenesis is a meiotic process
D. spermatogenesis is a mitotic process, while oogenesis is a meiotic process
E. spermatogenesis produces 4 1N sperms, while oogenesis produces 1 egg cell and polar bodies

24. How many Barr bodies are present in the white blood cells of a 48, XXYY individual?

A. 0
B. 1
C. 2
D. 3
E. 4

25. Polar bodies are the products of:

A. meiosis in females
B. meiosis in males
C. mitosis in females
D. mitosis in males
E. two of the above

26. Turner syndrome results from:

A. extra chromosome number 13
B. absence of the Y chromosome
C. presence of an extra Y chromosome
D. trisomy of the X chromosome
E. monosomy of an X chromosome

27. A sheep in 1996 named Dolly was different from animals produced by sexual reproduction because:

A. her DNA is identical to the DNA of her offspring

B. she was carried by a surrogate mother, while her DNA came from two other individuals

C. all her cells' DNA are identical

D. her DNA was taken from a somatic cell of an adult individual

E. her DNA was taken from a gamete of a neonatal individual

Notes for active learning

Development

1. Which statement is true regarding respiratory exchange during fetal life?

A. Respiratory exchanges are made through the placenta
B. Since lungs develop later in gestation, a fetus does not need a mechanism for respiratory exchange
C. The respiratory exchange is made through the ductus arteriosus
D. The respiratory exchange is not necessary
E. The respiratory exchange is made through the ductus venosus

2. When each cell in an early stage of embryonic development still has the ability to develop into a complete organism, it is known as:

A. gastrulation
B. blastulation
C. determinate cleavage
D. indeterminate cleavage
E. none of the above

3. Mutations in *Drosophila*, resulting in the transformation of one body segment into another, effect:

A. maternal-effect genes
B. homeotic genes
C. execution genes
D. segmentation genes
E. gastrulation genes

4. Which structure is developed from embryonic ectoderm?

A. connective tissue
B. bone
C. rib cartilage
D. epithelium of the digestive system
E. hair

5. The placenta is a vital metabolic organ made from a contribution by the mother and fetus. The portion of the placenta contributed by the fetus is:

A. amnion
B. yolk sac
C. umbilicus
D. chorion
E. none of the above

6. What occurs when an embryo lacks human chorionic gonadotropin (hCG) synthesis?

A. The embryo does not support the maintenance of the corpus luteum
B. The embryo increases the production of progesterone
C. The embryo develops immunotolerance
D. The placenta forms prematurely
E. Conception does not occur

7. Which germ layer gives rise to smooth muscle?

A. ectoderm
B. epidermis
C. mesoderm
D. endoderm
E. hypodermis

8. In early embryogenesis, the most critical morphological change in the development of cellular layers is:

A. epigenesis
B. gastrulation
C. conversion of morula to blastula
D. acrosomal reaction
E. fertilization membrane

9. During the first eight weeks of development, all events occur EXCEPT:

A. myelination of the spinal cord
B. presence of all body systems
C. formation of a functional cardiovascular system
D. beginning of ossification
E. limb buds appear

10. Which structures derive from the ectoderm germ layers?

A. blood vessels, tooth enamel, and epidermis
B. heart, kidneys, and blood vessels
C. nails, epidermis, and blood vessels
D. epidermis and adrenal cortex
E. epidermis and neurons

11. A diagram of the blastoderm that identifies regions from which specific adult structures are derived is a:

A. phylogenetic tree
B. pedigree diagram
C. fate map
D. linkage map
E. lineage diagram

12. Which tissue is the precursor of long bones in the embryo?

A. hyaline cartilage
B. fibrocartilage
C. dense fibrous connective tissue
D. elastic cartilage
E. costal cartilage

13. What changes are observed in cells as development proceeds?

A. Cytoskeletal elements involved in forming the mitotic spindle
B. Energy requirements of each cell
C. Genetic information that was duplicated with each round of cell division
D. Composition of polypeptides within the cytoplasm
E. Composition of nucleotides within the nucleus

14. During fertilization, what is the earliest event in the process?

A. sperm contacts the cortical granules around the egg
B. sperm nucleic acid enters the egg's cytoplasm to form a pronucleus
C. acrosome releases hydrolytic enzymes
D. sperm contacts the vitelline membrane around the egg
E. hyaluronic acid forms a barrier to prevent polyspermy

15. The embryonic ectoderm layer gives rise to all the following EXCEPT:

A. eyes
B. fingernails
C. integument
D. nervous system
E. blood vessels

16. All are correct matches of a fetal structure with what it becomes at birth, EXCEPT:

A. ductus venosus—ligamentum venosum
B. umbilical arteries—medial umbilical ligament
C. foramen ovale—fossa ovalis
D. ductus arteriosus—ligamentum teres
E. all choices are correct

17. All statements regarding gastrulation are true, EXCEPT:

A. the primitive gut that results from gastrulation is the archenteron
B. for amphibians, gastrulation is initiated at the gray crescent
C. after gastrulation, the embryo consists of two germ layers of endoderm and ectoderm
D. for amphibians, blastula cells migrate during gastrulation by an invagination region of the blastopore
E. the mesoderm develops as the third primary germ layer during gastrulation

18. A homeobox is a:

A. protein involved in the control of meiosis
B. transcriptional activator of other genes
C. DNA binding site
D. sequence responsible for post-transcriptional modifications
E. sequence that encodes for a DNA binding motif

19. After the gastrulation's completion, the embryo undergoes the following:

A. cleavage
B. blastulation
C. blastocoel formation
D. neurulation
E. dedifferentiation

20. Shortly after implantation:

A. trophoblast forms two distinct layers
B. embryo undergoes gastrulation within 3 days
C. maternal blood sinuses bathe the inner cell mass
D. myometrial cells cover and seal off the blastocyst
E. the umbilical cord forms within 48 hours

21. What is the anatomical connection between the placenta and embryo?

A. chorion
B. umbilical cord
C. endometrium
D. corpus luteum
E. vas deferens

22. Which statement correctly illustrates the principle of induction during vertebrate development?

A. Neural tube develops into the brain, the spinal cord, and the nervous system
B. Secretion of TSH stimulates the release of thyroxine hormone
C. Ectoderm develops into the nervous system
D. Neurons synapse with other neurons via neurotransmitters
E. Presence of a notochord beneath the ectoderm results in the formation of a neural tube

23. The acrosomal reaction by the sperm is:

A. a transient voltage change across the vitelline envelope
B. the jelly coat blocking penetration by multiple sperms
C. the consumption of yolk protein
D. hydrolytic enzymes degrading the plasma membrane
E. the inactivation of the sperm acrosome

24. Which primary germ layer gives rise to the cardiovascular system, bones, and skeletal muscles?

A. endoderm
B. blastula
C. mesoderm
D. ectoderm
E. gastrula

25. The dorsal surface cells of the inner cell mass form:

A. notochord
B. placenta
C. one of the fetal membranes
D. structure called the embryonic disc
E. primitive streak

26. All occur in a newborn immediately after birth, EXCEPT:

A. the infant completely stops producing fetal hemoglobin
B. resistance in the pulmonary arteries decreases
C. pressure in the left atrium increases
D. pressure in both the inferior vena cava and the right atrium increase
E. ductus arteriosus constricts

27. Homeobox sequences are present in:

A. introns
B. exons
C. 3'-untranslated regions
D. 5'-untranslated regions
E. exon-intron boundaries

28. After 30 hours of incubation, the ectoderm tissue within the gastrula differentiates into different specific tissues, which supports the conclusion that:

I. cells become either endoderm or mesoderm
II. cells contain a different genome from their parental cells
III. gene expression is altered

A. III only
B. I and III only
C. I only
D. I and II only
E. II and III only

29. What is the function of the yolk sac in humans?

A. forms into the placenta
B. secretes progesterone in the fetus
C. gives rise to blood cells and gamete-forming cells
D. stores embryonic waste
E. stores nutrients for the embryo

30. The trophoblast is mostly responsible for forming:

A. placental tissue
B. lining of the endometrium
C. allantois
D. archenteron
E. chorion

31. Certain lesions to the mesodermal embryonic primary germ layer may stimulate the development of *spina bifida*, a congenital fissure in the lower vertebrae. Besides the spinal column, what other structures would be affected by such lesions?

I. intestinal epithelium
II. skin and hair
III. blood vessels
IV. muscles

A. I and II only
B. II and III only
C. III and IV only
D. I, III and IV only
E. IV only

32. The homeotic genes encode for:

A. repressor proteins
B. transcriptional activator proteins
C. helicase proteins
D. single-strand binding proteins
E. restriction enzymes

33. Which cells give rise to muscles in a frog embryo?

A. neural tube
B. ectoderm
C. endoderm
D. mesoderm
E. notochord

34. Which is NOT involved in the implantation of the blastocyst?

A. Settling of the blastocyst onto the prepared uterine lining
B. Adherence of the trophoblast cells to the uterine lining
C. Phagocytosis by the trophoblast cells
D. Inner cell mass giving rise to primitive streak
E. Endometrium proteolytic enzymes produced by the trophoblast cells

35. During labor, which hormone stimulates contractions of uterine smooth muscle?

A. oxytocin
B. prolactin
C. luteinizing hormone
D. hCG
E. estrogen

36. In a chick embryo, specific ectoderm cells give rise to wing feathers, while others develop into thigh feathers or foot claws. The ectodermal cells that develop into wing feathers were transplanted to an area that develops into thigh feathers or feet claws. It was observed that the transplanted cells developed into claws. Which best explains the experimental results?

A. ectoderm cells possess positional information
B. destiny of the cells was already determined
C. ectoderm cells can develop into any tissue
D. the underlying mesoderm-induced cells
E. ectoderm released growth factors

37. The *slow block* to polyspermy is due to the following:

A. transient voltage changes across membranes
B. jelly coat blocking sperm penetration
C. consumption of yolk protein
D. formation of the fertilization envelope
E. inactivation of the sperm acrosome

38. Polyspermy in humans results in:

A. mitotic insufficiency
B. interruption of meiosis
C. nonviable zygote
D. multiple births
E. formation of multiple placentas

39. Mesoderm gives rise to:

A. intestinal mucosa
B. nerves
C. skin
D. lung epithelium
E. heart

40. Genomic imprinting is:

A. inactivation of a gene by interruption of its coding sequence
B. organization of molecules in the cytoplasm to provide positional information
C. DNA modification in gametogenesis that affects gene expression in the zygote
D. suppression of a mutant phenotype because of a mutation in a different gene
E. mechanism by which enhancers distant from the promoter can still regulate transcription

41. Comparing a developing frog embryo and an adult, cells of which have a greater rate of translation?

A. Adult, because ribosomal production is more efficient in a mature organism
B. Adult, because a mature organism has more complex metabolic requirements
C. Embryo, because a developing organism requires more protein production than an adult
D. Embryo, because ribosomal production is not yet under regulatory control by DNA
E. Embryo, because the proteins must undergo more extensive post-translation processing

42. Which structure is the first to form during fertilization in humans?

A. blastula
B. morula
C. neural tube
D. ectoderm
E. zygote

43. It is impossible for sperm to be functional (i.e., able to fertilize the egg) until after:

I. they undergo capacitation
II. the tail disappears
III. they have been in the uterus for several days
IV. they become spermatids

A. I only
B. II and IV only
C. I and IV only
D. I and III only
E. IV only

44. Where is the deformity if a teratogen affects the endoderm development shortly after gastrulation?

A. nervous system
B. lens of the eye
C. liver
D. skeleton
E. connective tissue

45. Mutations that cause cells to undergo developmental fates of other cell types are:

A. heterochronic mutations
B. loss-of-function mutants
C. transection mutations
D. execution mutations
E. homeotic mutations

46. Each is true for the cells of an early gastrula's eye field, EXCEPT that they are:

A. terminally differentiated
B. competent
C. derived from the ectoderm layer
D. capable of becoming other ectoderm structures
E. undifferentiated ectoderm

47. What is the role of proteases and acrosin enzymes in reproduction?

A. They degrade the nucleus of the egg and allow the sperm to enter
B. They degrade the protective barriers around the egg and allow the sperm to penetrate
C. They direct the sperm to the egg through chemotaxis messengers
D. They neutralize the mucous secretions of the uterine mucosa
E. They degrade the protective barriers around the sperm and allow the egg to penetrate

48. When does the human blastocyst implant in the uterine wall?

A. about a week past fertilization
B. at blastulation
C. a few hours past fertilization
D. at primary germ formation
E. at primitive streak formation

49. Which primary layer develops into the retina of the eye?

I. endoderm II. ectoderm III. mesoderm

A. I only
B. II only
C. III only
D. II and III
E. I and III

50. What changes must occur in a newborn's cardiovascular system after the infant takes its first breath?

A. Ductus arteriosus constricts and is converted to the ligamentum arteriosum
B. The urinary system is activated at birth
C. Ductus venosus is severed from the umbilical cord, and visceral blood enters the vena cava
D. Foramen ovale between the atria of the fetal heart closes at the moment of birth
E. Foramen ovale becomes the medial umbilical ligament

51. Early activation of the mammalian zygote nucleus may be necessary because:

A. most developmental decisions are made under the influence of the paternal genome
B. mammalian oocytes are too small to store molecules needed to support the cleavage divisions
C. in gametogenesis, specific genes undergo imprinting
D. mammals do not have maternal-effect genes
E. none of the above

52. In reptiles, the aquatic environment necessary for the embryonic development of amphibians is replaced by:

A. use of lungs instead of gills
B. humid atmospheric conditions
C. shells which prevent the escape of gas
D. intrauterine development
E. amniotic fluid

53. What would be expected to form from a portion of cells destined to become the heart if it was excised from an early gastrula and placed in a culture medium?

A. undifferentiated mesoderm
B. undifferentiated ectoderm
C. differentiated endoderm
D. undifferentiated endoderm
E. differentiated ectoderm

54. During gestation, when can an ultrasound determine the sex of the fetus?

A. at the midpoint of the first trimester
B. about 18 weeks after fertilization
C. at the end of the first trimester
D. at the midpoint of the second trimester
E. at the midpoint of the third trimester

55. How long is the egg viable and able to be fertilized after ovulation?

A. 36-72 hours
B. a full week
C. up to 6 hours
D. 24-36 hours
E. 12-24 hours

56. Which stage of embryonic development has a hollow ball of cells surrounding a fluid-filled center?

A. 3-layered gastrula
B. blastula
C. morula
D. zygote
E. 2-layered gastrula

57. Which statement describes the acrosome of a sperm cell?

A. It contains nucleic acid
B. It contains hydrolytic enzymes, which are released when the sperm encounters the jelly coat of the egg
C. It fuses with the egg's cortical granules
D. It functions to prevent polyspermy
E. It is used for the motility of the sperm along the Fallopian tubes (oviducts)

58. Which statement about fertilization is correct?

A. Most sperm cells are protected and remain viable once inside the uterus
B. If estrogen is present, the pathway through the cervical opening is blocked from sperm entry
C. The vagina's acidic environment destroys millions of sperm cells
D. Spermatozoa remain viable for about 72 hours in the female reproductive tract
E. The ovulated secondary oocyte is viable for about 72 hours in the female reproductive tract

59. All statements are correct regarding cleavage in human embryos, EXCEPT:

A. blastomeres are genetically identical to the zygote
B. holoblastic cleavage occurs in one portion of the egg
C. morula is a solid mass of cells produced via cleavage of the zygote
D. size of the embryo remains constant throughout the initial zygote cleavage
E. none of the above

60. All statements are correct regarding cleavage in human embryos, EXCEPT:

A. blastomeres are genetically identical to the zygote
B. holoblastic cleavage occurs in one portion of the egg
C. morula is a solid mass of cells produced via cleavage of the zygote
D. size of the embryo remains constant throughout the cleavage of the zygote
E. none of the above

Notes for active learning

DNA and Protein Synthesis

1. DNA and RNA differ because:

A. only DNA contains phosphodiester bonds
B. only RNA contains pyrimidines
C. DNA is in the nucleus, and RNA is in the cytosol
D. RNA is associated with ribosomes, and DNA is associated with histones
E. RNA contains a phosphate group in its ribose ring

2. In the 1920s, circumstantial evidence indicated that DNA was the genetic material. Which experiments led to the acceptance of this hypothesis?

A. Griffith's experiments with *Streptococcus pneumoniae*
B. Avery, MacLeod and McCarty's work with isolating the transforming principle
C. Hershey and Chase's experiments with viruses and radioisotopes
D. A, B and C were used to support this hypothesis
E. Darwin's theory of natural selection

3. The N-glycosidic bond is relatively unstable within a guanine molecule and can be hydrolyzed through depurination. Which molecule is likely to undergo depurination?

A. sterols
B. lipids
C. phospholipids
D. proteins
E. DNA

4. What process duplicates a single gene?

A. Unequal recombination at repeated sequences that flank the gene
B. Equal recombination at repeated sequences that flank the gene
C. Unequal recombination within a single gene
D. Equal recombination within a single gene
E. All the above

5. Which element is NOT within nucleic acids?

A. nitrogen
B. oxygen
C. phosphorus
D. sulfur
E. carbon

6. Which RNA molecule is translated?

A. miRNA
B. tRNA
C. rRNA
D. mRNA
E. C and D

7. The aging of normal cells is associated with:

A. loss of telomerase activity
B. a decrease in contact inhibition
C. an increase in mutation rate
D. activation of the maturation-promoting factor
E. extranuclear inheritance

8. Protein synthesis in eukaryotic cells initiates in which structures?

A. nucleus
B. Golgi
C. cytoplasm
D. rough endoplasmic reticulum
E. smooth endoplasmic reticulum

9. When a gene is duplicated on one chromatid, the gene on the other chromatid is:

A. duplicated
B. inverted
C. transposed to another site
D. maintained as a single gene
E. deleted

10. Experiments designed by Avery, McLeod, and McCarty to identify the transforming principle were based on:

I. purifying each of the macromolecule types from a cell-free extract
II. removing each of the macromolecules from a cell, then testing its type
III. selectively destroying the different macromolecules in a cell-free extract

A. I only
B. II only
C. III only
D. I, II and III
E. I and II only

11. What is the term for a blotting method where proteins are transferred from a gel to membranes and probed by antibodies to specific proteins?

A. Eastern blotting
B. Western blotting
C. Northern blotting
D. Southern blotting
E. both Northern and Western blotting

12. The figure shows a nucleotide. At what position will the incoming nucleotide be attached in the figure?

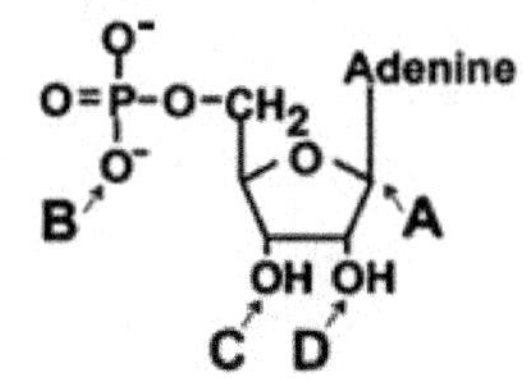

A. position A
B. position B
C. position C
D. position D
E. none of the above

13. All are correct about DNA, EXCEPT:

A. the strands are anti-parallel
B. the basic unit is nucleotide
C. the sugar molecule is deoxyribose
D. guanine binds to cytosine via three hydrogen bonds
E. adenine and guanine are pyrimidines

14. Tumor-suppressor genes normally control:

A. cell differentiation
B. necrosis
C. cell proliferation or activation of apoptosis
D. sister chromatid separation
E. protein degradation

15. Select the correct statement for aminoacyl tRNA synthetase.

A. It binds several different amino acids
B. It is an enzyme that uses energy from ATP to attach a specific amino acid to a tRNA
C. It is a tRNA that covalently binds amino acids
D. It synthesizes tRNA
E. It synthesizes rRNA

16. Griffith's experiment with pneumococcus demonstrated that:

A. smooth bacteria can survive heating
B. DNA, not protein, is the genetic molecule
C. materials from dead organisms can affect and change living organisms
D. nonliving viruses can change living cells
E. the virus injects its DNA into the host cell

17. The genetic code deciphered by Noble laureate Marshall W. Nirenberg (1927-2020) in 1964 encodes each amino acid by three nucleotides (codons). How many possible codons exist in nature that encode 20 amino acids in polypeptides?

A. 4
B. 20
C. 27
D. 64
E. 16

18. What mechanism targets proteins to organelles (e.g., chloroplast, mitochondrion)?

A. Addition of phosphate groups to the protein
B. Synthesizing the proteins as zymogens
C. Adding prosthetic groups to the protein
D. Cysteine bond formation
E. The signal sequence at the N-terminus of the polypeptide

19. When DNA is treated with 2-aminopurine, adenine is replaced by guanine on one strand. During replication, the complementary strand will have a substitution of:

A. guanine for adenine
B. adenine for guanine
C. cytosine for thymine
D. thymine for cytosine
E. adenine for cytosine

20. Before Nobel laureate Marshall W. Nirenberg *et al.* in 1964 determined the genetic code experimentally, why was it hypothesized that each codon would contain at least three bases?

A. Three bases are needed to produce a stable codon structure
B. There were three known nucleotide bases
C. There were more proteins than nucleotide bases
D. Three bases can form $4^3 = 64$ pairs, which is enough to encode 20 amino acids
E. There were twenty known amino acids

21. DNA of bacteria grown in a heavy (^{15}N) medium was isolated and added to an *in vitro* synthesis system. Then the bacteria are grown in a light (^{14}N) medium. After several hours, a sample of DNA was taken and analyzed for differing densities. How many DNA densities were in the sample after 2 generations?

A. 1
B. 2
C. 4
D. 8
E. 12

22. Which experimental procedure(s) simultaneously measure(s) the level of all mRNAs in a tissue?

I. Northern blot II. *In situ* hybridization III. Microarray experiment

A. I only
B. II only
C. III only
D. I, II and III
E. I and III only

23. To demonstrate that DNA is the "transforming principle," Avery, MacLeod, and McCarty showed that DNA could transform nonvirulent strains of pneumococcus. Their hypothesis was strengthened by their demonstration:

A. enzymes that destroyed proteins destroyed transforming activity
B. enzymes that destroyed nucleic acids destroyed transforming activity
C. enzymes that destroyed complex carbohydrates destroyed transforming activity
D. the transforming activity was destroyed by boiling
E. other strains of bacteria were transformed successfully

24. Which components of codon-anticodon hybridization on ribosomes determine the fidelity of protein synthesis?

A. mRNA & tRNA
B. mRNA & rRNA
C. tRNA & rRNA
D. DNA & RNA polymerase
E. RNA polymerase

25. Which procedure measures mRNA levels from only a single gene?

I. Northern blot II. *In situ* hybridization III. Microarray experiment

A. II only
B. I and II only
C. II and III only
D. I, II and III
E. I and III only

26. Which stage of cell division is the stage when chromosomes replicate?

A. prophase
B. telophase
C. anaphase
D. metaphase
E. interphase

27. If the transcript's sequence is 5'-CUAAGGGCUAC-3', what is the sequence of the DNA template?

A. 3'-GUAGCCCUUAG-5'
B. 3'-GTACGCCTTAG-5'
C. 5'-GTAACCCTTAG-3'
D. 5'-GUTACCUGUAG-3'
E. 5'-GTAGCCCTTAG-3'

28. Duplicated genes:

A. are more common in prokaryote genomes than in eukaryote genomes
B. are closely related but diverged in sequence and function over evolutionary time
C. never encodes for essential proteins, such as transcription factors
D. encode for proteins that catalyze different steps of a biochemical pathway
E. all the above

29. The Hershey–Chase experiment:

A. proved that DNA replication is semiconservative
B. used ^{32}P to label protein
C. used ^{35}S to label DNA
D. supported the hypothesis that DNA is the transforming molecule
E. both A and C

30. Which statement is NOT correct about DNA replication?

A. DNA polymerase synthesizes and proofreads the DNA
B. RNA primers are necessary for the hybridization of the polymerase
C. Ligase relaxes positive supercoils that accumulate as the replication fork opens
D. DNA polymerase adds Okazaki fragments in a 5' → 3' direction
E. DNA polymerase adds deoxynucleotides in a 5' → 3' direction

31. Which structures represent a peptide bond between adjacent amino acids?

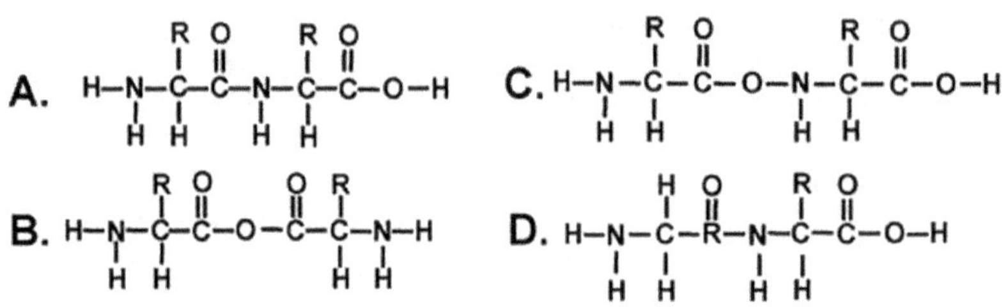

A. structure A
B. structure B
C. structure C
D. structure D
E. structures A and D

32. All these statements apply to proteins, EXCEPT:

A. they regulate cell membrane trafficking
B. they catalyze chemical reactions
C. they can be hormones
D. they undergo self-replication
E. they bind antigens

33. Eukaryote RNA polymerase usually:

A. binds to the TATAA promoter sequence and initiates transcription
B. needs general transcription factors to bind to the promoter and initiate basal-level transcription
C. needs specific regulatory transcription factors to bind to the promoter and initiate basal-level transcription
D. transcribes tRNA genes
E. transcribes mRNA genes

34. If an RNA sequence has a cytosine content of 25%, what is its adenine content?

A. 50%
B. 37.5%
C. 12.5%
D. 25%
E. cannot be determined

35. If a portion of prokaryotic mRNA has the base sequence 5′-ACUACUA$\underline{\text{U}}$GCGUCGA-3′, what could result from a mutation where the underlined base is changed to A?

I. truncation of the polypeptide
II. inhibition of initiation of translation
III. no effect on protein synthesis

A. I and II only
B. I and III only
C. II and III only
D. III only
E. II only

36. Which statement is INCORRECT about the genetic code?

A. Many amino acids are specified by more than one codon
B. Most codons specify more than one amino acid
C. There are multiple stop codons
D. Codons are 3 bases in length
E. The start codon inserts methionine at the amino end of the polypeptide

37. In bacteria, the enzyme that removes the RNA primers is:

A. DNA ligase
B. primase
C. reverse transcriptase
D. DNA polymerase I
E. helicase

38. Okazaki fragments are:

A. synthesized in a 5'→ 3' direction by DNA polymerase I
B. covalently linked by DNA polymerase I
C. components of the leading strand
D. components of DNA synthesized to fill in gaps after excision of the RNA primer
E. synthesized in a 5'→ 3' direction by DNA polymerase III

39. Peptide bond synthesis is catalyzed by:

A. tRNA in the cytoplasm
B. ribosomal proteins
C. ribosomal RNA
D. mRNA in the ribosome
E. none of the above

40. Which statement does NOT apply to protein synthesis?

A. The process does not require energy
B. rRNA is required for proper binding of the mRNA message
C. tRNA molecules shuttle amino acids assembled into polypeptides
D. The amino acid is bound to the 3' end of the tRNA
E. The mRNA is synthesized from 5' → 3'

41. All statements about PCR are correct, EXCEPT:

A. PCR can be used to obtain large quantities of a particular DNA sequence
B. PCR does not require knowledge of the terminal DNA sequences of the region to be amplified
C. PCR uses a DNA polymerase to synthesize DNA
D. PCR uses short synthetic oligonucleotide primers
E. PCR involves heating the DNA sample to denature complementary base pairing

42. The shape of a tRNA is determined primarily by:

A. its number of bases
B. proteins that bind it
C. tRNA and aminoacyl tRNA synthetase interactions
D. intramolecular base pairing
E. hydrophobic interactions

43. In prokaryotic cells, methylated guanine contributes to:

A. increased rate of DNA replication
B. decreased rate of DNA replication
C. correcting the separation of DNA strands
D. proofreading the replicated strands
E. correcting mismatched pairs of bases

44. In the polymerization reaction by DNA polymerase, what is the function of magnesium?

$$\text{Primer (free 3' OH)} + \text{5' PPP} \xrightarrow{Mg^{+2},\ 4\ \text{dNTPs, DNA polymerase}} \text{Primer 3'O-P-5'} + \text{PPi}$$

A. cofactor
B. monovalent metal ion
C. substrate
D. enzyme
E. coenzyme

45. The structure of the ribosome is created by:

I. internal base pairing of rRNA
II. ribosomal proteins
III. internal base pairing of mRNA
IV. internal base pairing of tRNA

A. I only
B. II only
C. I and II only
D. I, II and IV only
E. I, II and III only

46. Which statement is true for tRNA?

A. It has some short double-stranded segments
B. It has a poly-A tail
C. It is produced in the nucleolus
D. It is a long molecule of RNA
E. It is the template for protein synthesis

47. Which chemical group is at the 5' end of a single polynucleotide strand?

A. diester group
B. purine base
C. hydroxyl group
D. phosphate group
E. nitrogen group

48. The drug aminoacyl-tRNA is an analog of puromycin. Both have an amino group capable of forming a peptide bond, but puromycin lacks a carboxyl group to form another peptide bond. What is the possible effect of adding puromycin to bacteria undergoing protein synthesis?

A. Inhibition of initiation of protein synthesis
B. Inhibition of entry of aminoacyl-tRNA into the P site during elongation
C. Inability to form a complete ribosome
D. Substitution of puromycin for another amino acid in the protein, yielding a normal-length protein
E. Termination of protein synthesis via covalent attachment of puromycin

49. In *E. coli* cells, DNA polymerase I:

I. synthesizes most of the Okazaki fragments
II. simultaneously copies both strands of DNA
III. degrades the RNA primer portion of Okazaki fragments

A. I only
B. II only
C. III only
D. I and III only
E. I and II only

50. During DNA synthesis, the error rate is on the order of one mismatched nucleotide per:

A. 100
B. 1,000
C. 10,000
D. 1,000,000
E. 10,000,000

51. All are contained within a molecule of DNA, EXCEPT:

A. nitrogenous bases
B. phosphodiester bonds
C. polypeptide bonds
D. deoxyribose sugars
E. phosphate groups

52. In *E. coli* cells, DNA polymerase III:

A. synthesizes most of the Okazaki fragments
B. removes the RNA primer
C. is the only DNA polymerase used by *E. coli* during replication
D. degrades the RNA portion of an Okazaki fragment
E. synthesizes DNA in the 3' to 5' direction

53. Which molecule belongs to a different chemical category than the others?

A. uracil
B. guanine
C. adenine
D. thymine
E. cysteine

54. The enzyme that cleaves DNA at the sequence-specific site is:

A. restriction endonuclease
B. exonuclease
C. DNA polymerase
D. ligase
E. integrase

55. *E. coli* RNA polymerase-initiated transcription and synthesized one phosphodiester bond. Which molecule shown is RNA polymerase made from?

A.

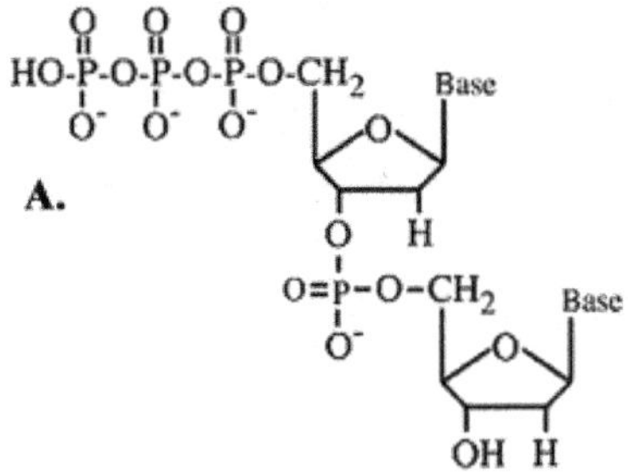

B.

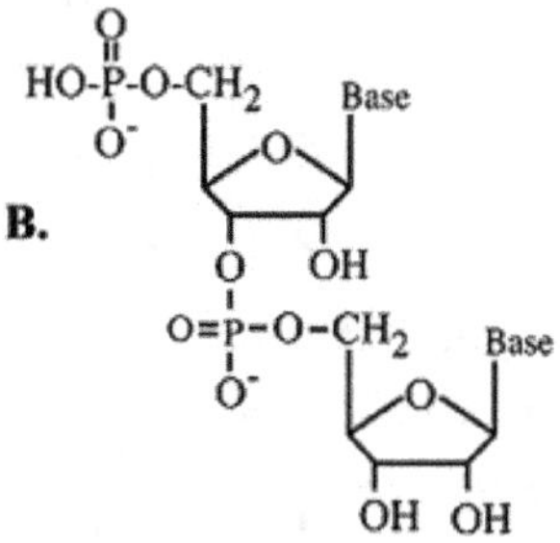

C.

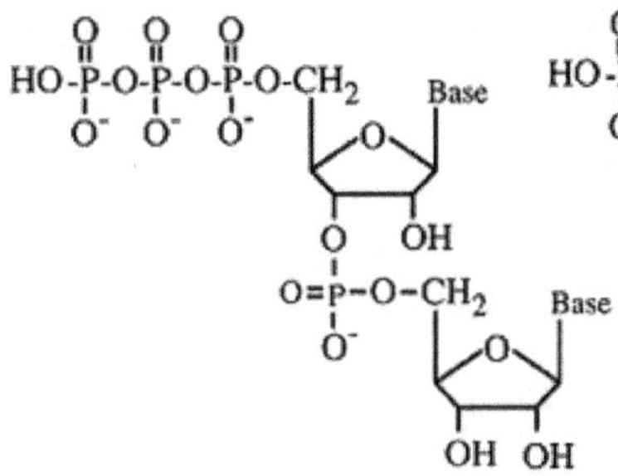

D.

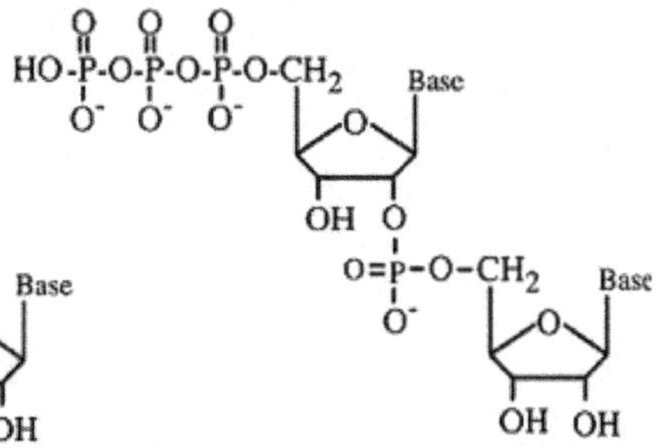

E.

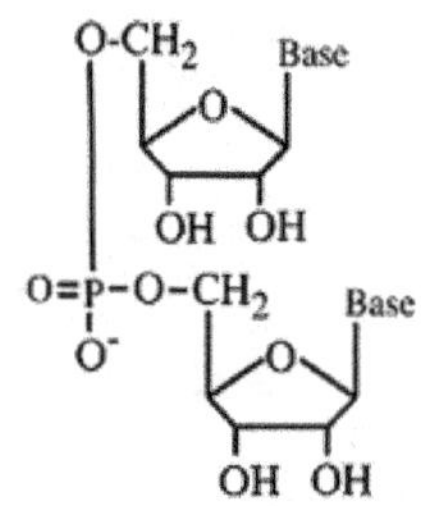

56. What rate does PCR increase the amount of DNA during each cycle?

A. additively
B. exponentially
C. linearly
D. systematically
E. gradually

57. Which is present in RNA but absent in DNA?

A. additional hydroxyl group
B. hydrogen bonds
C. thymine
D. double helix
E. phosphodiester bonds

58. After DNA strands are synthesized, which enzyme completes the process of DNA replication?

A. primase
B. ligase
C. helicase
D. reverse transcriptase
E. both B and D

59. When a base is paired with its complementary strand, which strand would have the highest melting point?

A. TTAGTCTC
B. TTTTAAAA
C. AGCTTCGT
D. CGCGTATA
E. GCCAGTCG

60. A technique that investigates gene function by mutating wildtype genes is:

A. contig building
B. transgenetics
C. reverse genetics
D. gene therapy
E. gene mapping

61. How many high-energy phosphate bonds are needed to translate a 50-amino acid polypeptide (starting with mRNA, tRNA, amino acids, and the necessary enzymes)?

A. 49
B. 50
C. 101
D. 199
E. 150

62. The mRNA in *E. coli* cells is composed primarily of:

A. four bases – A, T, C, G
B. phosphodiester linkages connecting deoxyribonucleotide molecules
C. two strands that base pair in an anti-parallel orientation
D. processed RNA molecule containing introns
E. phosphodiester linkages connecting ribonucleotide molecules

63. Which statement about DNA mismatch repair is correct?

A. DNA is scanned for any base-pairing mismatches after methyl groups are added to guanines
B. Errors in replication made by DNA polymerase are corrected on the unmethylated strand
C. The proofreading mechanism removes all abnormal bases
D. Repairs from high-energy radiation damage are made
E. Mismatch repair occurs on each strand of DNA during replication

64. What is the first amino acid of each protein of eukaryotic cells?

A. methionine
B. glutamate
C. valine
D. proline
E. isoleucine

65. DNA in *E. coli* is composed of:

I. four bases – A, T, C, G
II. phosphodiester linkages that connect deoxyribonucleotide molecules
III. two strands that base pair in an anti-parallel orientation
IV. phosphodiester linkages that utilize the 3'-OH

A. I and II only
B. I and III only
C. I, II and III only
D. I, II, III and IV
E. II, III and IV only

66. If a peptide has the sequence val-ser-met-pro and the tRNA molecules used in its synthesis have the corresponding sequence of anticodons 3'-CAG-5', 3'-UCG-5', 3'-UAC-5', 3'-UUU-5', what sequence of the DNA codes for this peptide?

A. 5'–CAGTCGTACTTT–3'
B. 5'–TTTCATGCTGAC–3'
C. 5'–GACGCTCATTTT–3'
D. 5'–UUUCAUGCUGAC–3'
E. 5'–CAGUCGUACUUU–3'

67. The site of the DNA template that RNA polymerase binds to during transcription is:

A. promoter
B. leader sequence
C. enhancer
D. domain
E. transcription factor

68. Which dipeptide is synthesized by a ribosome?

A. isoleucine-glycine
B. cytosine-guanine
C. proline-thymine
D. uracil-glutamic acid
E. isoleucine-glycine and uracil-glutamic acid

69. Which is the correct order of events in delivering a protein to its cellular destination?

A. Signal sequence binds to docking protein → transmembrane-gated channel opens → protein enters the organelle
B. Membrane channel is formed → signal sequence binds to docking protein → chaperonins unfold protein → protein enters the organelle → protein refolds
C. Chaperonins unfold protein → signal sequence binds to docking protein → membrane channel is formed → protein enters the organelle → protein refolds
D. Membrane channel is formed → chaperonins unfold protein → signal sequence binds to docking protein → protein enters the organelle → protein refolds
E. Signal sequence binds to docking protein → membrane channel is formed → chaperonins unfold protein → protein enters the organelle → protein refolds

70. Select the correct mRNA sequences depending on the direction RNA polymerase transcribes.

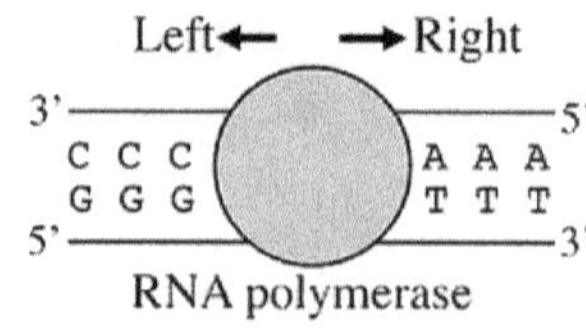

RNA sequence if RNA polymerase goes left:	RNA sequence if RNA polymerase goes right:
A. 5'-GGG-3'	5'-AAA-3'
B. 5'-GGG-3'	5'-UUU-3'
C. 5'-CCC-3'	5'-AAA-3'
D. 5'-CCC-3'	5'-UUU-3'
E. None of the above	

71. Which statement is TRUE for the base composition of DNA?

A. In double-stranded DNA, the number of G bases equals the number of T bases
B. In double-stranded DNA, the number of A bases equals the number of T bases
C. In double-stranded DNA, the number of C bases equals the number of T bases
D. In every single strand, the number of A bases equals the number of T bases
E. In double-stranded DNA, the number of G bases equals the number of A bases

72. The figure shows a replication fork in *E. coli*. Which of the indicated sites is the 3' end of the lagging strand?

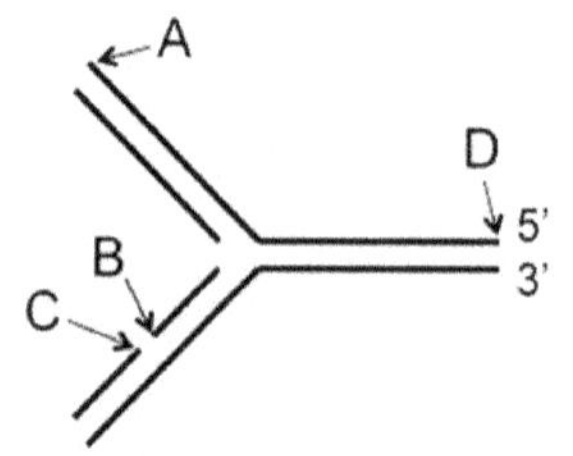

A. site A
B. site B
C. site C
D. site D
E. sites C and D

73. Ribosomal subunits are isolated from bacteria grown in a "heavy" ^{13}C and ^{15}N medium and added to an *in vitro* system that actively synthesizes protein. Following translation, a sample is removed and centrifuged. Which would be the best illustration of centrifugation results?

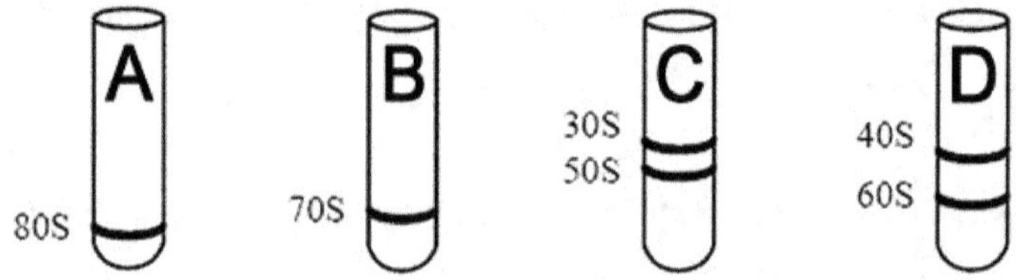

A. Test tube A
B. Test tube B
C. Test tube C
D. Test tube D
E. Test tubes A and C

74. Which statement is TRUE?

A. polypeptides are synthesized by the addition of amino acids to the amino terminus
B. prokaryotic RNA usually undergoes nuclear processing
C. RNA polymerase has a proof-reading activity
D. prokaryotic RNA contains introns
E. 3' end of mRNA corresponds to the carboxyl terminus of the protein

75. A codon for histidine is 5'-CAU-3'. The anticodon in the tRNA that brings histidine to the ribosome is:

A. 5'-CAU-3'
B. 5'-GUA-3'
C. 5'-UAC-3'
D. 5'-AUG-3'
E. none of the above

76. During translation elongation, the existing polypeptide chain is transferred to which site as the ribosome moves in the 3' direction?

A. tRNA occupying the A site
B. tRNA occupying the P site
C. ribosomal rRNA
D. signal recognition particle
E. none of the above

77. Which primer will amplify the following DNA fragments *via* PCR?

5′-ATCGGTATGTAACGCTCACCTGT-3′

A. 5′-ACAG-3′
B. 5′-AGAC-3′
C. 5′-TAGC-3′
D. 5′-GACT-3′
E. 5′-CTGT-3′

78. Which statement about the genetic code is FALSE?

A. It is mostly the same for *E. coli* and humans
B. It is redundant
C. It is ambiguous
D. It has one codon for starting translation
E. All the above are true

79. What portion of the polypeptide chain is responsible for establishing and maintaining the force used to stabilize the secondary structure?

A. C-terminus
B. N-terminus
C. carbonyl oxygen
D. R-groups
E. both A and B

First base (5' end)	U		C		A		G		Third base (3' end)
U	UUU	Phe	UCU	Ser	UAU	Tyr	UGU	Cys	U
	UUC		UCC		UAC		UGC		C
	UUA	Leu	UCA		UAA	Stop	UGA	Stop	A
	UUG		UCG		UAG	Stop	UGG	Trp	G
C	CUU	Leu	CCU	Pro	CAU	His	CGU	Arg	U
	CUC		CCC		CAC		CGC		C
	CUA		CCA		CAA	Gln	CGA		A
	CUG		CCG		CAG		CGG		G
A	AUU	Ile	ACU	Thr	AAU	Asn	AGU	Ser	U
	AUC		ACC		AAC		AGC		C
	AUA		ACA		AAA	Lys	AGA	Arg	A
	AUG	Met start	ACG		AAG		AGG		G
G	GUU	Val	GCU	Ala	GAU	Asp	GGU	Gly	U
	GUC		GCC		GAC		GGC		C
	GUA		GCA		GAA	Glu	GGA		A
	GUG		GCG		GAG		GGG		G

Second base: U, C, A, G

80. A ribosome has made a tripeptide, MET-ARG-SER, attached to the tRNA in the P site. Using the genetic code table, what codon is in the E site of the ribosome?

A. AUG
B. CGU
C. UCA
D. UGA
E. It cannot be determined

81. A ribosome has made a tripeptide, MET-ARG-SER, attached to the tRNA in the P site. Using the genetic code table, what codon is in the A site of the ribosome?

A. AUG
B. CGU
C. UCA
D. UGA
E. It cannot be determined

82. Which is an example of a transversion mutation where purine gets converted to pyrimidine?

A. uracil → thymine
B. cytosine → thymine
C. thymine → adenine
D. guanine → adenine
E. guanine → cytosine

83. Recombinant DNA experiments utilize plasmids because:

A. they contain foreign DNA
B. their genetic material cannot be cut with restriction enzymes
C. they are unable to replicate inside the bacteria
D. they are used to transform bacteria
E. they are a natural part of the bacterial genome

84. Which task requires recombinant DNA technology?

A. Creating bacteria that produce human growth hormone
B. Creating a polyploid cherry tree
C. Crossing two types of orange trees to create new orange fruits
D. Crossing a donkey with a horse to breed a mule
E. Paternity testing

85. Gene therapy is successful when:

A. the person's cells express the newly introduced gene
B. the replacement gene is integrated into viral DNA
C. the virus with the replacement gene enters the person's cells
D. the person's cells replicate the newly introduced gene
E. the replacement gene is integrated into the person's genome

Notes for active learning

Genetics and Inheritance Patterns

1. Which characteristic makes an organism unsuitable for genetic studies?

A. Large number of chromosomes
B. Short generation time
C. Ease of cultivation
D. Ability to control crosses
E. Availability of a variation for traits

2. People with the sex-linked genetic disease hemophilia suffer from excessive bleeding because their blood will not clot. Tom, Mary, and their four daughters do not exhibit symptoms of hemophilia. However, their son exhibits symptoms of hemophilia because:

A. Tom is heterozygous
B. Tom is homozygous
C. Mary is heterozygous
D. Mary is homozygous
E. All the above are equally probable

3. Several eye colors are characteristic of *Drosophila melanogaster*. Red eyes are dominant over sepia or white eyes. What percent of offspring of a sepia-eyed fly will have sepia eyes if mated with a red-eyed fly that was a cross of red-eyed and sepia-eyed parents?

A. 0%
B. 25%
C. 50%
D. 75%
E. 100%

4. Color blindness mutations in humans result from:

A. fragile X syndrome
B. chromosome nondisjunction
C. reciprocal translocation
D. dosage compensation
E. unequal crossing-over

5. Which method was NOT used by Mendel to study the genetics of garden peas?

A. Maintenance of true-breeding lines
B. Cross-pollination
C. Microscopy
D. Production of hybrid plants
E. Quantitative analysis of results

6. Crossing AAbbCc × AaBbCc where A, B and C are unlinked genes, what is the probability of obtaining offspring with the AaBbCc genotype?

A. 1/4
B. 1/16
C. 1/64
D. 1/32
E. 1/8

7. What is the probability of having a child affected by a disease with an autosomal recessive inheritance if the mother and father are carriers of the disease?

A. 0%
B. 25%
C. 50%
D. 75%
E. 66%

8. For the multi-step progression of cancer, the major mutational target(s) is/are:

A. telomerase
B. X-linked traits
C. tumor suppressor gene
D. trinucleotide repeats
E. transcription factors

9. Since the gene responsible for color blindness is on the X chromosome, what is the probability that a son of a colorblind man and a woman carrier will be colorblind?

A. 75%
B. 100%
C. 25%
D. 50%
E. 66%

10. If two strains of true-breeding plants with alleles for a specific character are crossed, their progeny are:

A. P generation
B. F_1 generation
C. F_2 generation
D. F_1 crosses
E. F_2 progeny

11. The Arabidopsis plant has five pairs of homologous chromosomes. Suppose an Arabidopsis is heterozygous for five mutations, and each mutation is on a different chromosome. How many genetically distinct gametes will this plant make after meiosis?

A. 5
B. 10
C. 32
D. 64
E. 25

12. An unknown inheritance pattern has the following characteristics:

- 25% probability of having a homozygous unaffected child
- 25% probability of having a homozygous affected child
- 50% probability of having a heterozygous child

Which Mendel's inheritance pattern best matches the above observations?

A. autosomal recessive
B. autosomal dominant
C. X-linked recessive
D. X-linked dominant
E. cannot be determined without more information

13. What is the frequency of heterozygotes within a population in Hardy-Weinberg equilibrium if the frequency of the dominant allele D is three times that of the recessive allele d?

A. 7.25%
B. 12.75%
C. 33%
D. 37.5%
E. 50%

14. A recessive allele may appear in a phenotype due to:

A. gain-of-function mutation
B. acquired dominance
C. senescence
D. processivity
E. the loss of heterozygosity

15. Which observations would support the theory of maternal inheritance for the spunky phenotype?

A. Spunky female x wild-type male → progeny all spunky
B. Wild-type female x spunky male → progeny all spunky
C. Wild-type female x spunky male → progeny 1/2 spunky, 1/2 wild-type
D. Spunky female x wild-type male → progeny 1/2 spunky, 1/2 wild-type
E. Spunky female x wild-type male → progeny all wild-type

16. Mendel concluded that each pea has two units for each characteristic, and each gamete contains one unit. Mendel's "unit" is now referred to as:

A. genome
B. hnRNA
C. codon
D. transcription factor
E. gene

17. Which leads to a complete loss of gene function?

A. A missense mutation that causes the nonpolar methionine to be replaced with glycine
B. GC base pair being converted to an AT base pair in the promoter
C. A mutation in the third codon of the open reading frame
D. A base pair change that does not affect the amino acid sequence
E. All the above

18. All effects are possible after a mutation, EXCEPT:

A. abnormal lipid production
B. abnormal protein production
C. gain of enzyme function
D. loss of enzyme function
E. no change in protein production

19. Which cross must produce all green, smooth peas if green (G) is dominant over yellow (g) and smooth (S) is dominant over wrinkled (s)?

A. GgSs × GGSS
B. GgSS × ggSS
C. Ggss × GGSs
D. GgSs × GgSs
E. None of the above

20. Retinoblastoma is inherited as:

A. a multifactorial trait
B. X-linked recessive
C. Mendelian dominant
D. Mendelian recessive
E. an extranuclear trait

21. Tay-Sachs disease is a rare autosomal recessive genetic disorder. If a male heterozygous carrier and a female heterozygous carrier have a first child who is homozygous wild type, what is the probability that the second child develops Tay-Sachs?

A. 1/3
B. 1/2
C. 1/16
D. 1/8
E. 1/4

22. Mendel's crossing of spherical-seeded pea plants with wrinkled-seeded pea plants resulted in progeny that all had spherical seeds. This indicates that the wrinkled-seed trait is:

A. codominant
B. dominant
C. recessive
D. penetrance
E. codominant and recessive

23. The result of mitosis is the production of:

A. two (1N) cells identical to the parent cell
B. two (2N) cells identical to the parent cell
C. four (1N) cells identical to the parent cell
D. four (2N) cells identical to the parent cell
E. four (1N) unique cells that are genetically different from the parent cell

24. The degree of genetic linkage is often measured by the:

A. frequency of nonsense mutations
B. histone distribution
C. frequency of missense mutations
D. probability of crossing over
E. AT/GC ratio

25. Cancers associated with defects in mismatch repair are inherited via:

A. dominant inheritance
B. maternal inheritance
C. X-linked inheritance
D. epigenetic inheritance
E. recessive inheritance

26. Given that color blindness is a recessive trait inherited through a sex-linked gene on the X chromosome, what probability will a daughter born to a colorblind father and a mother carrier be a carrier?

A. 0%
B. 25%
C. 50%
D. 100%
E. 12.5%

27. What is the probability that a cross between a true-breeding pea plant with a dominant trait and a true-breeding pea plant with a recessive trait will result in all F_1 progeny having the dominant trait?

A. 50%
B. 25%
C. 0%
D. 100%
E. 12.5%

28. The tall allele is dominant to short. True-breeding tall plants were crossed with true-breeding short plants. The F_1 plants were self-crossed to produce F_2 progeny. What are the phenotypes of the F_1 and F_2 progeny?

A. All F_1 and 1/4 of the F_2 are short
B. All F_1 are short, and 1/4 of the F_2 are tall
C. All F_1 and 3/4 of the F_2 are tall
D. All F_1 are tall, and 3/4 of the F_2 plants are short
E. All the above are equally probable

29. All DNA lesions result in a frameshift mutation, EXCEPT:

A. 1 inserted base pair
B. 2 substituted base pairs
C. 4 inserted base pairs
D. 2 deleted base pairs
E. 5 deleted base pairs

30. If two species with the AaBbCc genotype reproduce, what is the probability that their progeny have the AABBCC genotype?

A. 1/2
B. 1/4
C. 1/16
D. 1/64
E. 1/8

31. At the hypoxanthine-guanine phosphoribosyltransferase (HPRT) locus, an average amount of mRNA is present, but no protein is observed. This phenotype is caused by the following:

A. frameshift mutation
B. mutation in the gene-altering the restriction pattern but not affecting the protein; for instance, the mutated nucleotide is in the third codon position of the open reading frame
C. point mutation leading to an amino acid substitution necessary for enzyme function
D. gene deletion or mutation affecting the promoter
E. nonsense mutation affecting message translation

32. Hemophilia is a recessive X-linked trait. Knowing that females with Turner's syndrome have a high incidence of hemophilia, it can be concluded that these females have:

A. lost an X and gained a Y
B. lost an X
C. gained an X
D. gained a Y
E. none of the above

33. What is the pattern of inheritance for a rare recessive allele?

A. Every affected person has an affected parent
B. Unaffected parents can produce children who are affected
C. Unaffected mothers have affected sons and daughters who are carriers
D. Every affected person produces an affected offspring
E. None of the above

34. True-breeding plants with large purple flowers were crossed with true-breeding plants with small white flowers. The F_1 progeny all had large purple flowers. The F_1 progeny were crossed to true-breeding plants with small white flowers. Among 1000 progeny:

Number of progeny	Flower size	Flower color
250	small	white
250	small	purple
250	large	white
250	large	purple

Most likely, the genes for flower size and color:

A. are unlinked
B. are sex-linked
C. are linked and separated by no more than 25 centimorgans
D. require determination of the cross of the F_2 progeny
E. cannot be determined

35. If tall height and brown eye color are dominant, what is the probability for a heterozygous tall, heterozygous, brown-eyed mother and a homozygous tall, homozygous, blue-eyed father to have a tall child with blue eyes? Note: the genes for eye color and height are unlinked.

A. 3/4
B. 1/8
C. 1/4
D. 1/2
E. None of the above

36. Recombination frequencies:

A. are the same for *cis*- and *trans*-heterozygotes
B. arise from completely random genetic exchanges
C. decrease with distance
D. are the same for all genes
E. are the same for all chromosomes

37. How many different gametes can be produced from the genotype AaBbCc, assuming independent assortment?

A. 4
B. 6
C. 8
D. 16
E. 3

38. What is the pattern of inheritance for a rare dominant allele?

A. Every affected person has an affected parent
B. Unaffected parents can produce children who are affected
C. Unaffected mothers have affected sons and daughters who are carriers
D. Every affected person produces an affected offspring
E. All the above

39. Individuals homozygous for an autosomal recessive mutation accumulate harmful amounts of lipids. Jane and her parents are not afflicted. However, Jane's sister accumulates lipids. What is the probability that Jane is heterozygous for the mutation?

A. 1/4
B. 1/3
C. 2/3
D. 1/2
E. 3/4

40. A genetic disease with early-onset and severe symptoms with every generation is an example of:

A. codominance
B. penetrance
C. heterozygous advantage
D. gain-of-function mutations
E. anticipation

41. For a trait with two alleles, if the recessive allele frequency is 0.6 in a population, what is the frequency of individuals expressing the dominant phenotype?

A. 0.48
B. 0.64
C. 0.16
D. 0.36
E. 0.12

42. The maximum recombination frequency between two genes is:

A. 100%
B. 80%
C. 50%
D. 10%
E. 1%

43. In mice, short hair is dominant over long hair. If a short-haired individual is crossed with a long-haired individual, and both long and short-haired offspring result, what can be concluded?

A. Short-haired individual is homozygous
B. Short-haired individual is heterozygous
C. Long-haired individual is homozygous
D. Long-haired individual is heterozygous
E. More information is required

44. The result of meiosis in males is the production of:

A. two (1N) cells genetically identical to the parent cell
B. two (2N) cells genetically identical to the parent cell
C. four (1N) cells genetically identical to the parent cell
D. four (1N) unique cells genetically different from the parent cell
E. four (2N) cells genetically identical to the parent cell

45. Which is a type of genetic mutation?

I. insertion II. frameshift III. nonsense IV. missense

A. I and II only
B. I, II and III only
C. II and IV only
D. I, II, III and IV
E. II, III and IV only

46. Two reciprocal crossing-over events appear in the progeny at an approximate ratio of:

A. 4:1
B. 3:1
C. 2:1
D. 2:3
E. 1:1

47. In dogs, phenotype A (erect ears and barking while following a scent) is caused by dominant alleles; recessive alleles cause phenotype B (droopy ears and silent while following a scent). A dog that is homozygous dominant for both traits is mated with a dog that is homozygous recessive for both traits. If the two genes are unlinked, which is the expected F_1 phenotypic ratio?

A. 9:3:3:1
B. 1:1
C. 16:0
D. 1:2:1
E. None of the above

48. Mutations:

A. always cause severe mutant phenotypes
B. never cause severe mutant phenotypes
C. are not inherited by the progeny
D. may cause premature termination of translation
E. are none of the above

49. What is the probability of having a child affected by a disease with autosomal dominant inheritance if both the mother and father have one mutant gene for that disease?

A. 0%
B. 25%
C. 50%
D. 12.5%
E. 75%

50. Given the recombinant frequencies below, what is the sequence of linked genes D, E, F and G?

GE: 23%	ED: 15%	EF: 8%
GD: 8%	GF: 15%	DF: 7%

A. FGDE
B. EFGD
C. GFDE
D. GDFE
E. DEFG

Questions **51** through **57** are based on the following:

The pedigree illustrated by the schematic shows the inheritance of albinism, a homozygous recessive condition manifested in a lack of pigment. Specify the genotypes using *A* and *a* to indicate dominant and recessive alleles.

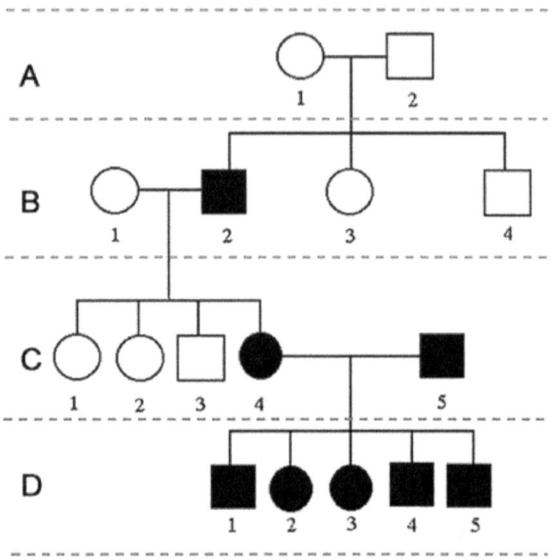

Note: solid figures are albino individuals.

51. Individual A-1 in the pedigree shown is:

A. *AA*
B. *aa*
C. *Aa*
D. any of the above
E. none of the above

52. Individual A-2 in the pedigree shown is:

A. *AA*
B. *aa*
C. *Aa*
D. any of the above
E. none of the above

53. Individual B-1 in the pedigree shown is:

A. *AA*
B. *aa*
C. *Aa*
D. any of the above
E. none of the above

54. Individual B-2 in the pedigree shown is:

A. *AA*
B. *aa*
C. *Aa*
D. any of the above
E. none of the above

55. Individual C-3 in the pedigree shown is:

A. *AA*
B. *aa*
C. *Aa*
D. any of the above
E. none of the above

56. Individual C-4 in the pedigree shown is:

A. *AA*
B. *aa*
C. *Aa*
D. any of the above
E. none of the above

57. Individual D-4 in the pedigree shown is:

A. *AA*
B. *aa*
C. *Aa*
D. any of the above
E. none of the above

58. In cocker spaniels, black color (B) is dominant over red (b), and solid color (S) is dominant over spotted (s). If the genes are unlinked and the offspring of BBss and bbss individuals are mated, what fraction of their offspring will be black and spotted?

A. 1/16
B. 9/16
C. 1/9
D. 3/16
E. 3/4

59. Why do genes that cause disease often appear to skip generations in an X-linked recessive inheritance?

A. The disease is primarily transmitted through unaffected carrier females
B. Males with an affected gene are carriers but do not show the disease
C. X-linked diseases are only expressed in males
D. All X-linked diseases display incomplete penetrance
E. none of the above

60. The "calico" coat pattern of a female cat is a result of:

A. endoreduplication
B. unequal crossing-over
C. random X chromosome inactivation
D. Turner syndrome
E. trisomy of the X chromosome

61. Which statement is true for an autosomal dominant inheritance?

I. A single allele of the mutant gene is needed to exhibit the phenotype
II. Transmission to the son by the father is not observed
III. Autosomal dominant traits do not skip generations

A. II only
B. I, II and III
C. I only
D. I and III only
E. II and III only

62. What chromosomal abnormality results in some XY individuals being phenotypically females?

A. fragile X syndrome
B. Barr body formation
C. dosage compensation
D. mosaicism
E. portion deleted of the Y chromosome with the testis-determining factor

63. Which event risks combining two recessive alleles, resulting in a genetic defect?

A. genetic mutation
B. transformation
C. inbreeding
D. crossing over
E. trisomy

64. To engineer polyploid plants in plant breeding, genetic engineers use drugs that:

A. insert new DNA into plants' genome
B. damage the DNA and cause mutations
C. rearrange the sequences of codons on the DNA strand
D. alter the number of chromosomes
E. shorten the length of DNA strands

65. To create mules, horses are bred with donkeys by:

A. crossing over
B. genetic engineering
C. hybridization
D. inbreeding
E. genetic mutation

66. When treated with penicillin, a bacterial culture transformed with recombinant plasmids that contain a gene for resistance to this antibiotic will:

A. undergo lysis
B. rapidly replicate DNA
C. die
D. survive
E. alternate generations

67. Which pairing represents two transgenic organisms?

A. Appletree hybrid and a polyploid cherry tree
B. Human growth hormone-producing bacteria and genetically modified soybeans
C. Appletree hybrid and human growth hormone-producing bacteria
D. Genetically modified soybeans and a polyploid cherry tree
E. Polyploid cherry tree and human growth hormone-producing bacteria

68. Which choice correctly describes the process of establishing parental relationships through DNA fingerprinting of specific genes?

A. Mitochondrial DNA links a daughter to the mother, while plasmid DNA links to the father
B. Mitochondrial DNA links a son to the mother, while the Y chromosome links to the father
C. X chromosome links a daughter to the mother, while the Y chromosome links to the father
D. X chromosome links a son to the mother, while mitochondrial DNA links to the father
E. Y chromosome links a son to the mother, while the X chromosome links a girl to the father

69. Which inheritance pattern is when an affected male has all affected daughters but no affected sons?

A. X-linked recessive
B. Y-linked
C. Autosomal dominant
D. Autosomal recessive
E. X-linked dominant

Notes for active learning

Cell & Molecular Biology

Answer Keys
&
Detailed Explanations

Page intentionally left blank

Answer Keys

Chapter 1. Eukaryotic Cell

1: D	11: C	21: C	31: B	41: D	51: B	61: B
2: A	12: E	22: B	32: D	42: C	52: A	62: D
3: E	13: B	23: E	33: E	43: C	53: B	63: A
4: C	14: C	24: A	34: B	44: A	54: D	64: E
5: E	15: E	25: B	35: C	45: B	55: C	65: B
6: C	16: C	26: E	36: D	46: E	56: E	66: A
7: A	17: D	27: C	37: E	47: D	57: A	67: E
8: A	18: A	28: D	38: C	48: E	58: C	68: D
9: B	19: B	29: A	39: D	49: B	59: A	69: E
10: B	20: A	30: B	40: B	50: D	60: E	

Chapter 2. Cellular Metabolism and Enzymes

1: D	11: A	21: A	31: E	41: B	51: C	61: A
2: A	12: D	22: B	32: C	42: C	52: B	62: D
3: C	13: E	23: A	33: D	43: A	53: C	63: E
4: B	14: D	24: E	34: B	44: E	54: D	
5: E	15: B	25: C	35: D	45: A	55: D	
6: C	16: D	26: D	36: B	46: C	56: D	
7: D	17: C	27: C	37: A	47: C	57: C	
8: A	18: B	28: E	38: D	48: B	58: C	
9: B	19: E	29: B	39: E	49: E	59: C	
10: E	20: B	30: B	40: A	50: B	60: B	

Chapter 3. Photosynthesis

1: E	11: C	21: A	31: C	41: A	51: A
2: D	12: C	22: B	32: A	42: B	52: B
3: A	13: E	23: E	33: E	43: B	53: D
4: B	14: B	24: C	34: C	44: C	54: C
5: C	15: A	25: B	35: A	45: E	55: B
6: D	16: A	26: E	36: D	46: A	56: E
7: A	17: B	27: D	37: E	47: D	57: C
8: B	18: E	28: D	38: B	48: A	
9: A	19: D	29: C	39: A	49: E	
10: E	20: A	30: A	40: B	50: C	

Chapter 4. Molecular Biology of Eukaryotes

1: A	11: A	21: A	31: C	41: A	51: A
2: E	12: B	22: C	32: A	42: E	52: E
3: C	13: E	23: B	33: E	43: B	53: B
4: A	14: C	24: C	34: B	44: D	54: B
5: C	15: D	25: A	35: A	45: E	
6: D	16: A	26: E	36: B	46: B	
7: C	17: B	27: D	37: E	47: C	
8: A	18: E	28: B	38: D	48: C	
9: E	19: C	29: E	39: D	49: A	
10: D	20: B	30: D	40: B	50: D	

Chapter 5. Biomolecules

1: A	11: B	21: A	31: A	41: D	51: C
2: B	12: B	22: B	32: E	42: C	52: C
3: A	13: D	23: A	33: C	43: D	53: C
4: B	14: B	24: B	34: D	44: D	
5: C	15: D	25: D	35: D	45: A	
6: B	16: C	26: D	36: D	46: B	
7: C	17: B	27: C	37: C	47: D	
8: C	18: B	28: D	38: A	48: A	
9: D	19: D	29: C	39: D	49: B	
10: B	20: B	30: B	40: B	50: A	

Chapter 6. Specialized Cells and Tissues

1: E	11: A	21: B	31: E	41: D	51: C
2: A	12: C	22: B	32: B	42: E	52: A
3: A	13: E	23: E	33: D	43: D	53: C
4: D	14: C	24: D	34: E	44: A	54: B
5: B	15: E	25: A	35: A	45: A	55: E
6: E	16: D	26: B	36: C	46: B	56: D
7: B	17: D	27: C	37: B	47: C	57: B
8: D	18: B	28: C	38: D	48: A	58: E
9: C	19: A	29: E	39: A	49: E	
10: D	20: B	30: D	40: C	50: D	

Chapter 7. Microbiology

1: E	11: A	21: A	31: C	41: C	51: B
2: B	12: B	22: D	32: E	42: E	52: E
3: D	13: E	23: E	33: D	43: C	53: C
4: D	14: C	24: B	34: B	44: A	54: E
5: B	15: E	25: E	35: D	45: C	55: A
6: C	16: E	26: D	36: E	46: A	56: D
7: D	17: B	27: C	37: A	47: A	57: C
8: A	18: C	28: B	38: B	48: B	58: A
9: C	19: A	29: A	39: A	49: E	59: E
10: E	20: C	30: C	40: C	50: B	60: C

Chapter 8. Mechanisms of Reproduction

1: B	11: E	21: D
2: C	12: D	22: A
3: E	13: C	23: E
4: C	14: D	24: B
5: C	15: D	25: A
6: D	16: B	26: E
7: C	17: E	27: D
8: B	18: C	
9: E	19: A	
10: B	20: E	

Chapter 9. Development

1: A	11: C	21: B	31: C	41: C	51: C
2: D	12: A	22: E	32: B	42: E	52: E
3: B	13: D	23: D	33: D	43: A	53: A
4: E	14: C	24: C	34: D	44: C	54: C
5: D	15: E	25: E	35: A	45: E	55: E
6: A	16: D	26: D	36: D	46: A	56: B
7: C	17: C	27: B	37: D	47: B	57: B
8: B	18: E	28: A	38: C	48: A	58: C
9: A	19: D	29: C	39: E	49: B	59: B
10: E	20: A	30: A	40: C	50: A	60: B

Chapter 10. DNA and Protein Synthesis

1: D	11: B	21: B	31: A	41: B	51: C	61: D	71: B	81: E
2: D	12: C	22: C	32: D	42: D	52: A	62: E	72: B	82: E
3: E	13: E	23: B	33: B	43: E	53: E	63: B	73: C	83: D
4: A	14: C	24: A	34: E	44: A	54: A	64: A	74: E	84: A
5: D	15: B	25: B	35: A	45: C	55: C	65: D	75: D	85: A
6: D	16: C	26: E	36: B	46: A	56: B	66: B	76: B	
7: A	17: D	27: E	37: D	47: D	57: A	67: A	77: A	
8: C	18: E	28: B	38: E	48: E	58: B	68: A	78: C	
9: E	19: C	29: D	39: C	49: C	59: E	69: E	79: C	
10: C	20: D	30: C	40: A	50: D	60: C	70: D	80: B	

Chapter 11. Genetics and Inheritance Patterns

1: A	11: C	21: E	31: E	41: B	51: C	61: D
2: C	12: A	22: C	32: B	42: C	52: C	62: E
3: C	13: D	23: B	33: B	43: B	53: C	63: C
4: E	14: E	24: D	34: A	44: D	54: B	64: D
5: C	15: A	25: A	35: D	45: D	55: C	65: C
6: E	16: E	26: C	36: A	46: E	56: B	66: D
7: B	17: B	27: D	37: C	47: C	57: B	67: B
8: C	18: A	28: C	38: A	48: D	58: E	68: B
9: D	19: A	29: B	39: C	49: E	59: A	69: E
10: B	20: C	30: D	40: E	50: D	60: C	

Section I: Cell Structure & Function

Eukaryotic Cell

1. D is correct.

Facilitated transport (or *passive transport*) uses a protein pore (or channel) but is differentiated from *active transport, which requires energy.*

2. A is correct.

Cell is the unit of function and reproduction because subcellular components cannot regenerate whole cells.

3. E is correct.

I: water relies on *aquaporins* to readily diffuse across a plasma membrane.

Without aquaporins, only a small fraction of water molecules can diffuse through the cell membrane per unit of time because of the polarity of water molecules.

II: *small hydrophobic molecules* readily diffuse through the hydrophobic tails of the plasma membrane.

III: *small ions* rely on ion channel transport proteins to diffuse across the membrane.

IV: *neutral gas molecules* (e.g., O_2, CO_2) readily diffuse through the hydrophobic tails of the plasma membrane.

4. C is correct.

Integral proteins often span the plasma membrane.

Plasma membrane is a phospholipid bilayer with an inner span of hydrophobic (or *water-fearing*) regions.

Diameter of a plasma membrane is 20 to 25 amino acids thick.

Spans of 20-25 hydrophobic residues prefer to be embedded in the plasma membrane and isolated from water.

Hydropathy analysis determines the degree of *hydrophobicity* (i.e., nonpolar or *water-fearing*) or *hydrophilicity* (i.e., polar or *water-loving*) of amino acids in a protein. It characterizes the possible structures of a protein.

Each residue (i.e., individual amino acid) has a hydrophobicity value (analogous to electronegativity).

Hydropathy analysis plots the degree of hydrophobicity (or *hydrophilicity*) on the y-axis and the amino acid sequence on the x-axis, *hydrophobicity vs. amino acid* position.

Amino acids have lower energy when polar amino acids occupy a polar environment (e.g., cytoplasm) and nonpolar amino acids occupy a nonpolar environment (e.g., the interior of the plasma membrane).

5. E is correct.

Lysosome is the digestive region of the cell and is a membrane-bound organelle with a low pH (around 5) that stores hydrolytic enzymes.

A: *vacuoles* and *vesicles* are membrane-bound sacs involved in the transport and storage of materials ingested, secreted, processed, or digested by cells.

Vacuoles are larger than vesicles and are in plant cells (e.g., central vacuole).

C: *chloroplasts* are the site of photosynthesis and are only in algae and plant cells.

Like mitochondria, chloroplasts contain circular DNA and ribosomes and may have evolved similarly via endosymbiosis.

D: *phagosomes* are vesicles for transporting and storing materials ingested by the cell through phagocytosis. Vesicles form by the fusion of the cell membrane around the particle.

Phagosome is a cellular compartment in which pathogenic microorganisms are digested.

Phagosomes *fuse with lysosomes* in their maturation process to form *phagolysosomes.*

6. C is correct.

DNA damage checkpoints are *signal transduction pathways* that block cell cycle progression in G1, G2, and metaphase and slow the S phase progression rate when DNA is damaged.

Pausing cell cycle allows the cell to repair the damage before dividing.

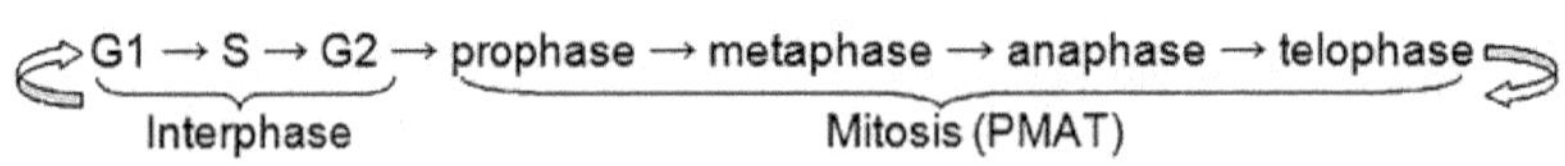

Cell cycle divides into interphase (G1. S, G2) and mitosis (prophase, metaphase, anaphase, and telophase)

7. A is correct.

(C≡G) base pairs are linked in the double helix by *three hydrogen bonds.*

(A=T) base pairs are linked in the double helix by *two hydrogen bonds.*

Therefore, it takes more energy to separate G-C base pairs.

Less G–C rich strands of double-stranded DNA require less energy to separate (i.e., denature).

Chargaff's rule specifies complementary base pairing; double-stranded DNA has equal G and C (and A and T).

8. A is correct.

Phospholipids are lipids as major components of cell membranes; they form lipid bilayers.

Phospholipids contain a glycerol backbone, a phosphate group, and a simple organic molecule (e.g., choline).

The 'head' is *hydrophilic* (i.e., attracted to water), while the 'tails' are *hydrophobic* (i.e., repelled by water), and the tails aggregate (via hydrophobic forces).

Hydrophilic heads contain negatively charged phosphate groups and glycerol.

Hydrophobic tails usually consist of 2 long fatty acids (saturated or unsaturated) hydrocarbon chains.

Cholesterol is embedded within animal lipid bilayers but is absent in plant cell membranes. Embedded cholesterol in the phospholipid bilayer in eukaryotic animal cells allows for protective cell membranes that can also change shape.

Plant and bacterial cell walls are primarily composed of cellulose and peptidoglycan glucose polymers, respectively, which are rigid and restrict movement.

9. B is correct.

Cytoskeleton determines the *overall shape* of a cell and is composed of:

Microtubules help synthesize cell walls in plants which primarily contribute to shape; centrosomes organize microtubules, cilia, and flagella (9+2).

Microtubules form *mitotic spindles* and *centrioles* (non-membrane-bound organelles).

Intermediate filaments: anchor organelles (e.g., nucleus) and bear tension, contributing to cell shape.

Microfilaments: resist compression, thus contributing to cell shape.

10. B is correct.

Osmolarity is determined by the total concentration of dissolved particles in the solution.

Compounds that dissociate into ions (i.e., electrolytes) increase the concentration of particles and produce a higher osmolarity.

To determine which molecule (after dissociation into ions) generates the highest osmolarity, determine the number of individual ions each molecule dissociates into in H_2O.

$CaCl_2$ dissociates into 1 Ca^{2+} and 2 Cl^-

As a result, $CaCl_2$ generates the greatest osmolarity, which equals 250 mOsmoles for Ca^{2+} + 500 mOsmoles for Cl^- = 750 mOsmoles when $CaCl_2$ dissociates into ions within the solution.

A: NaCl dissociates in water with an osmolarity of 600 mOsmoles because it dissociates into 1 Na^+ and 1 Cl^-.

Na^+ cations and Cl^- anions are added: 300 mOsmoles for Na^+ + 300 mOsmoles for Cl^- = 600 mOsmoles.

C: glucose does *not* dissociate in water and has the same osmolarity as the starting molecule—500 mOsmoles.

D: urea does *not* dissociate in water and has the same osmolarity as the starting molecule— 600 mOsmoles.

11. C is correct.

The two ribosomal subunits are synthesized in the nucleolus, a region within the nucleus. The ribosomes are the sites of protein production.

Prokaryotic ribosomes (30S small + 50S large subunit = 70S complete ribosome) are smaller than *eukaryotic ribosomes* (40S small + 60S large subunit = 80S complete ribosome).

A: *Golgi apparatus* is a membrane-bound organelle that modifies (e.g., glycosylation), sorts, and packages proteins synthesized by the ribosomes.

B: *lysosomes* have a low pH of about 5 and contain hydrolytic enzymes involved in digestion.

D: *rough endoplasmic reticulum* (RER) is part of the endomembrane extending from the nuclear envelope.

RER has ribosomes associated with its membrane and is the site of the production and folding of proteins.

Misfolded proteins exit the rough ER and are sent to the *proteasome* for degradation.

E: *cell membrane* is a barrier between the interior and exterior of the cell.

12. E is correct.

Cyclins are phosphorylated proteins responsible for specific events during cycle division, such as microtubule formation and chromatin remodeling.

Cyclins are four classes based on their behavior in the cell cycle: G1/S, S, M, and G1 cyclins.

p53 is not a transcription factor but is a tumor suppressor gene. The p53 protein is crucial in multicellular organisms, where it regulates the cell cycle and functions as a tumor suppressor (preventing cancer).

p53 is *the guardian of the genome* because it conserves stability by preventing genome mutation.

13. B is correct.

Codon is a three-nucleotide segment of an mRNA that hybridizes (via complementary base pairing) with the appropriate anticodon on the tRNA to encode one amino acid in a polypeptide chain during protein synthesis.

The tRNA molecule interacts with the mRNA codon after the ribosomal complex binds the mRNA.

A: tRNA molecules interact with the mRNA codon after (not before) the ribosomal complex binds mRNA

C: *translation* involves the conversion of mRNA into protein.

D: *operons* regulate the transcription of genes into mRNA and are not involved in translating mRNA into proteins.

14. C is correct.

Peroxisomes are organelles in most eukaryotic cells; a significant function is the breakdown of very-long-chain fatty acids through *beta-oxidation.*

Peroxisomes convert the very long fatty acids to medium-chain fatty acids in animal cells, subsequently shuttled to the mitochondria. Medium-chain fatty acids are degraded, via oxidation, into CO_2 and H_2O.

Peroxisomes are membrane-bound, while ribosomes are non-membrane-bound organelles.

15. E is correct.

Vacuole is a membrane-bound organelle in plant and fungal cells and some protist, animal, and bacterial cells.

Vacuoles are enclosed compartments filled with water. They contain inorganic and organic molecules (including enzymes in solution) and may contain solids that have been engulfed.

Function and significance of vacuoles vary by cell type, with much greater prominence in plants, fungi, and certain protists than in animals or bacteria.

Mitochondria and chloroplasts have circular DNA resembling DNA in prokaryotes, as supported by the endosymbiotic *theory* for the evolution of mitochondria and chloroplasts from prokaryotes.

Ribosomes are the site of protein synthesis.

16. C is correct.

I-cell disease patients cannot correctly direct newly synthesized peptides to their target organelles. The Golgi apparatus is part of the endomembrane system and serves as a cellular distribution center.

Golgi apparatus packages proteins before they are sent to their destination. It modifies, sorts, and packages proteins for cell secretion (i.e., exocytosis) or use within the cell.

A: *nucleus* is the largest organelle, contains genetic material (i.e., DNA), and is the site of rRNA synthesis (within the nucleolus).

D: *smooth ER* (endoplasmic reticulum) forms the endomembrane system, is connected to the nuclear envelope, and functions in several metabolic processes. It synthesizes lipids, phospholipids, and steroids.

Cells secreting lipids, phospholipids, and steroids (e.g., testes, ovaries, and skin oil glands) have an extensive smooth endoplasmic reticulum.

Smooth ER conducts the metabolism of carbohydrates and drug detoxification.

It is responsible for the attachment of receptors on cell membrane proteins and steroid metabolism.

E: *nucleolus* is the organelle within the nucleus responsible for synthesizing ribosomal RNA (rRNA).

17. D is correct.

Cytochrome c oxidase is a large transmembrane protein complex in bacteria and the mitochondrion of eukaryotes. It receives an electron from each of four cytochrome c molecules and transfers the electrons to an O_2 molecule, converting molecular oxygen to two molecules of H_2O.

Cytochrome c oxidase is the *last enzyme in the electron transport chain* (ETC) of mitochondria (or bacteria) in the mitochondrial (or bacterial) inner membrane.

Mitochondria and chloroplasts have circular DNA resembling DNA in prokaryotes.

Endosymbiotic theory illustrates the evolution of mitochondria and chloroplasts from prokaryotes, thus explaining the resemblance.

18. A is correct.

Cytoskeleton is integral to proper cell division because it forms the mitotic spindle and separates sister chromatids during cell division.

Cytoskeleton comprises microtubules and microfilaments, provides mechanical cell support to maintain shape, and functions in cell motility.

19. B is correct.

Biochemical events during the cell cycle include DNA damage repair and replication completion, centrosome duplication, spindle assembly, and attachment of the kinetochores to the spindle.

20. A is correct.

Mitochondria divide autonomously (i.e., independent of the genome) to produce daughter mitochondria that incorporate new nonradioactive phosphatidylcholine and inherit radioactive phosphatidylcholine from the parent via *semiconservative replication.*

Therefore, the daughter mitochondria have equal radioactivity.

B: mitochondria divide *autonomously*, and daughter mitochondria retain parental (original) radioactive label.

C: original sample was 100% radiolabeled.

DNA replication is *semiconservative*, whereby one strand of parental DNA (radiolabeled) and one strand of newly replicated (non-radiolabeled) are in each daughter cell after the first round of division.

D: requires the daughter mitochondria to be synthesized *de novo* (i.e., new) with newly synthesized, nonradioactive phosphatidylcholine.

If the mitochondria divide autonomously, the daughter mitochondria retain the radioactive label evenly via *semiconservative replication.*

21. C is correct.

Golgi apparatus (i.e., Golgi complex) is a eukaryotic cell organelle.

Golgi apparatus processes proteins via *post-translational modifications* for three destinations:

1) *secreted* from the cell,

2) transported into *organelles,* or

3) targeted to the *plasma membrane.*

Golgi complex mainly processes proteins synthesized by the ER

In plant cells, the central vacuole functions as a lysosome, stores nutrients and maintains osmotic balance.

Peroxisomes are like lysosomes in size, are bound by a single membrane, and are filled with enzymes.

However, peroxisomes *bud from* the endoplasmic reticulum.

22. B is correct.

Prokaryotic cells (bacteria) have a typical cell width of 0.2 to 2.0 micrometers in diameter.

Eukaryotic cells (animal cells) have a typical cell width of 10-100 micrometers in diameter.

1 millimeter = 1×10^{-3} m

1 micrometer = 1×10^{-6} m

1 nanometer = 1×10^{-9} m

Light microscopes visualize objects from 1 millimeter (10^{-3} m) to 0.2 micrometers (2×10^{-7} m).

Electron microscopes visualize objects as small as an atom (1 angstrom or 10^{-10} m).

Microscopic scale ranges from 1 millimeter (10^{-3} m) to a ten-millionth of a millimeter (10^{-10} m). There are immense variations in objects' sizes, even within the microscopic scale.

10^{-3} m is 10 million times larger than 10^{-10} m, equivalent to the Earth's size *vs.* a beach ball.

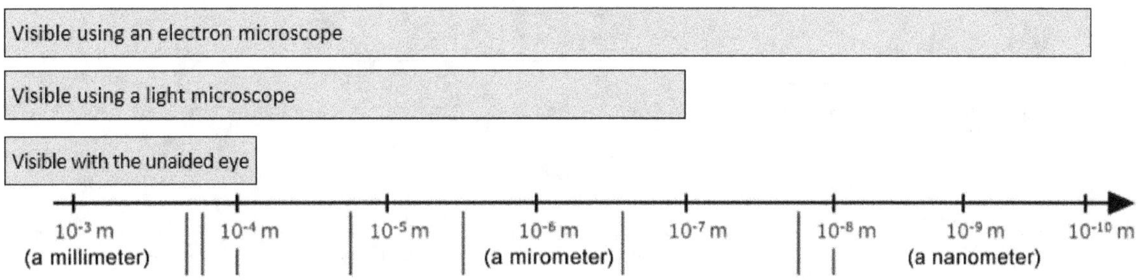

Comparison of resolution by unaided eyes, light, and electron microscopes

23. E is correct.

Ribosomes are composed of specific rRNA molecules and associated proteins.

Ribosomes are identified by the sedimentation coefficients (i.e., S units for Svedberg units) for density.

Prokaryotes have a 30S small and a 50S large subunit (i.e., complete ribosome = 70S; based on density).

Eukaryotes have a 40S small and 60S large subunit (i.e., complete ribosome = 80S).

A: *peroxisomes* are organelles involved in hydrogen peroxide (H_2O_2) synthesis and degradation. They function in cell detoxification and contain the catalase that decomposes H_2O_2 into H_2O and O_2.

C: *mitochondria* are organelles as the site of cellular respiration (i.e., oxidation of glucose to yield ATP) and plentiful in cells with high demands for ATP (e.g., muscle cells).

The number of mitochondria within a cell varies widely by organism and tissue type.

Many cells have a single mitochondrion, whereas others contain several thousand mitochondria.

Nucleus is the largest membrane-bound organelle in eukaryotes, containing the genetic code (i.e., DNA).

It directs the storing and transmitting of genetic information. Cells can contain multiple nuclei (e.g., skeletal muscle cells), one nucleus, or none (e.g., red blood cells).

24. A is correct.

Osmosis is a type of diffusion involving water and is a form of passive transport.

Hypertonic means a solution of high solute and low solvent concentrations.

Hypotonic means a solution of high solvent and low solute concentrations.

Solvents flow spontaneously from an area of high solvent to a low solvent concentration.

During *osmosis*, water flows from a hypotonic to a hypertonic environment.

25. B is correct.

Complex of Cdk and cyclin B is a maturation or mitosis-promoting factor (MPF).

Cyclin B is necessary to progress cells into and out of the M phase of the cell cycle.

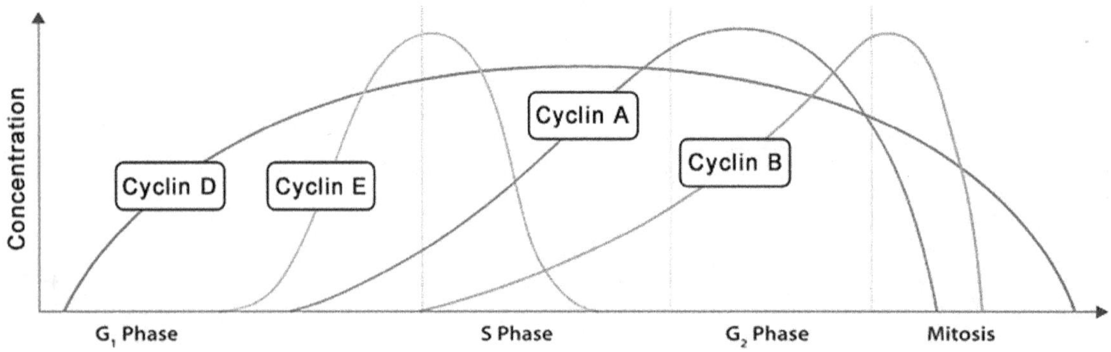

26. E is correct.

Mitochondria have their DNA genetic material and machinery to manufacture their RNAs and proteins.

For example, the trait is recessive (not observed in limited number of offspring) and encoded by a nuclear gene.

A: like organisms that reproduce sexually, mice inherit the *mitochondrial organelle* from their mother and display maternal inheritance of mitochondrial genes.

C: if X-linked, selectively male (not female) progeny would display the trait.

D: mitochondrial genes cannot be *recessive* because mitochondria are *inherited from the mother*.

27. C is correct.

Smooth endoplasmic reticulum (smooth ER) participates in synthesizing phosphatidylcholine.

Smooth ER functions include:

1) synthesis of lipids, phospholipids, and steroids

2) metabolism of carbohydrates and steroids

3) detoxifying alcohol and drugs

4) regulating Ca^{2+} concentration in muscle cells

Phosphatidylcholine is a class of phospholipids.

28. D is correct.

Plastids (e.g., chloroplast and chromoplast) are major organelles in plants and algae.

Plastids are the site of manufacturing and storing critical chemical compounds used by cells.

They often contain pigments used in photosynthesis. Pigments change to determine cell color.

Plastids, like prokaryotes, contain a circular double-stranded DNA molecule.

29. A is correct.

Thyroid gland synthesizes calcitonin in response to high blood calcium levels. It acts to reduce blood calcium (Ca^{2+}), opposing the effects of parathyroid hormone (PTH).

Calcitonin lowers blood Ca^{2+} levels in three ways:

1) inhibiting Ca^{2+} absorption by the intestines,

2) inhibiting osteoclast activity in bones, and

3) inhibiting renal tubular cell reabsorption of Ca^{2+}, allowing it to be excreted in the urine.

B: *kidneys* serve several essential regulatory roles. They are essential in the urinary system and serve homeostatic functions such as regulating electrolytes, maintaining acid-base balance, and regulating blood pressure (maintaining salt and water balance).

Kidneys secrete *renin* (involved in blood pressure regulation) that induces the release of aldosterone from the adrenal cortex (it increases blood pressure via sodium reabsorption, increasing blood pressure).

C: *parathyroid glands* synthesize parathyroid hormone (PTH), which increases blood calcium.

PTH increases calcium concentration in blood by acting upon the parathyroid hormone receptor (elevated levels in bone and kidney) and the parathyroid hormone receptor (elevated levels in the central nervous system, pancreas, testes, and placenta).

D: *liver* is the largest organ with many functions, including detoxification, protein synthesis, and the production of biomolecules necessary for digestion. The liver synthesizes bile, which is necessary for dietary lipid emulsification (in the small intestine).

E: *spleen* is a *reservoir* for red blood cells and *filters* the blood.

30. B is correct.

Rough endoplasmic reticulum (rough ER) participates in synthesizing plasma membrane proteins.

Oxidative phosphorylation is a series of redox reactions in the electron transport chain (ETC), leading to the production of ATP. In Eukaryotes, this occurs in the *mitochondrial inner membrane* and in prokaryotes in the intermembrane space.

Endocytosis is when a cell membrane invaginates, forming a vacuole to store molecules from the extracellular space actively transported across the cell's plasma membrane.

Post-translational modification is after ribosomes translate mRNA into polypeptide chains; the polypeptide chains become a mature protein by undergoing biochemical reactions (e.g., cleavage, folding).

31. B is correct.

Active transport uses a carrier protein and energy to move a substance across a membrane against (i.e., up) a concentration gradient: from low solute to a region of high solute concentration.

Donnan equilibrium refers to some ionic species passing through the barrier while others cannot.

Charged substances that cannot pass through the membrane create an uneven electrical charge.

Donnan potential is the electric potential arising between the solutions.

32. D is correct.

Cyclin D synthesis is initiated during G1 and drives the G1/S phase transition.

Apoptosis is programmed cell death (PCD) that may occur in multicellular organisms. Biochemical events lead to characteristic cell changes (morphology) and death.

Apoptotic changes include blebbing (i.e., an irregular bulge in the plasma membrane), cell shrinkage, chromatin condensation, nuclear fragmentation, and chromosomal DNA fragmentation.

In contrast to necrosis, traumatic cell death resulting from acute cellular injury, apoptosis confers advantages during an organism's life cycle.

For example, the differentiation of fingers and toes in a developing human embryo occurs because cells between the fingers undergo apoptosis, and the digits are separated.

Unlike necrosis, apoptosis produces cell fragments called *apoptotic bodies* that phagocytic cells can engulf and quickly remove before the cell's contents can spill out onto surrounding cells and cause damage.

33. E is correct.

During meiosis I, *homologous chromosomes* separate.

Sister chromatids (identical copies, except for recombination) separate during meiosis II.

Klinefelter syndrome (XXY karyotype) contains two X and one Y chromosome.

X and Y would be "homologous chromosomes" and typically separate during meiosis I.

Failure to separate during meiosis I could create a sperm containing an X and a Y, causing Klinefelter syndrome.

Anaphase is when the *centromere splits*, and the *homologous chromosomes / sister chromatids* are drawn away (via spindle fibers) from each other toward opposite sides of the two cells.

Homologous chromosomes separate during anaphase I, while sister chromatids separate during anaphase II.

In females, Turner's syndrome is due to the single X karyotype (single X chromosome and lacking a Y).

34. B is correct.

Most proteins that are secretory, membrane-bound, or targeted to an organelle use the *N-terminal signal sequence* (i.e., 5 to 30 amino acids) to target the protein.

Signal sequence of the polypeptide is recognized by a signal recognition particle (SRP), while the protein is synthesized on ribosomes.

Synthesis pauses while the ribosome-protein complex transfers to an SRP receptor on the ER (in eukaryotes) or plasma membranes (in prokaryotes) before polypeptide translation resumes.

35. C is correct.

Ribosomes anchored to the endoplasmic reticulum create the *rough endoplasmic reticulum* "studded" with ribosomes in contrast to the *smooth endoplasmic reticulum*, which lacks ribosomes.

"*Free*" and "*attached*" ribosome comparisons:

1) *free ribosomes* are in the cytoplasm, while *attached ribosomes* are anchored to the endoplasmic reticulum (ER), and

2) *free ribosomes* produce proteins in the cytosol, while *attached ribosomes* produce proteins inserted into the ER lumen (i.e., interior space).

36. D is correct.

Anterior pituitary hormones (including GH) are peptides.

Peptide hormones are *hydrophilic* and *cannot* cross the hydrophobic phospholipid bilayer; peptide hormones bind to receptors on the plasma membrane.

Destruction of the plasma membrane dramatically *reduces* the concentration of GH receptors.

37. E is correct.

Urea, a byproduct of amino acid metabolism, is a small uncharged molecule that crosses cell membranes by simple diffusion – a passive process that does not require energy.

A and C: export of Na^+ from a neuron is coupled with the *import of* K^+; the sodium-potassium-ATPase pump is an ATP-dependent process necessary to maintain a voltage potential across the neuron membrane.

B: movement of Ca^{2+} into a muscle cell occurs against the concentration gradient, and Ca^{2+} enters cells by *active transport*, which requires ATP.

D: *synaptic vesicles* contain neurotransmitters, and their exocytosis at a nerve terminus is an ATP-dependent process triggered by an action potential propagating along the neuron.

Vesicle fusion requires Ca^{2+} to enter the cell upon axon depolarization reaching the terminus.

38. C is correct.

p53 is a *tumor suppressor protein* that regulates the cell cycle and prevents cancer in multicellular organisms.

p53 is *the guardian of the genome* because of its role in conserving genetic stability by inhibiting genome mutations. The name p53 refers to its apparent molecular mass of 53 kDa.

Dalton is defined as 1/12 the mass of carbon and is a unit of a convention for expressing (i.e., often in kilodaltons or kDa) the molecular mass of proteins.

39. D is correct.

Newly synthesized secretory protein pathway:

rough ER → Golgi → secretory vesicles → exterior of the cell (via exocytosis)

Peroxisomes and *lysosomes* are destinations for proteins but do not involve secretory path exocytosis.

Ribosomes synthesize proteins, but the Golgi is the final organelle before exocytosis.

40. B is correct.

Plant cell membranes have higher amounts of unsaturated fatty acids.

The ratio of saturated and unsaturated fatty acids determines membrane fluidity.

Unsaturated fatty acids have kinks in their tails (due to double bonds) that push the phospholipids apart so the membrane retains its fluidity.

A: some textbooks state that plant membranes lack cholesterol, but a small amount is present, which is negligible compared to animal cells.

Cholesterol is usually dispersed in varying amounts throughout animal cell membranes in the irregular spaces between the hydrophobic lipid tails of the membrane.

Cholesterol functions as a *bidirectional buffer.*

At *elevated temperatures*, cholesterol *decreases membrane fluidity* because it confers stiffening and strengthening effects on the membrane.

At *low temperatures*, cholesterol *intercalates* between the phospholipids, preventing clustering and *stiffening the membrane.*

41. D is correct.

Nucleus, chloroplast, and mitochondria are organelles enclosed in a double membrane.

Endosymbiotic theory illustrates the evolution of mitochondria and chloroplasts from prokaryotes, thus explaining the resemblance.

42. C is correct.

Golgi apparatus processes *secretory proteins* via post-translational modifications.

Proteins targeted to the Golgi (from the rough ER) have three destinations:

1) secreted out of the cell,
2) transported into organelles, or
3) targeted to the plasma membrane (as a receptor, channel, or pore).

43. C is correct.

Carbon, hydrogen, and oxygen are in macromolecules (i.e., proteins, carbohydrates, nucleic acids, and lipids).

Nitrogen is in nucleic acids and proteins, which may contain sulfur or phosphorus.

Amino acids are *monomers* (building blocks) of proteins, and *nitrogen* is an *amino acid* and *urea* component.

Organic nitrogen is a nitrogen compound originating from a living organism.

Carbohydrates and lipids are made of only carbon, hydrogen, and oxygen

Carbohydrates have about 2 H and 1 O atom for every C atom.

Proteins are composed of C, H, O, N, and sometimes S.

Nucleic Acids are composed of: nucleotides composed of C, H, O, N, and P.

44. A is correct.

Albumin is the most abundant plasma protein; primarily responsible for *osmotic pressure* in circulatory systems.

Albumin is too large to pass from the circulatory system into the interstitial space.

Osmotic pressure is the force of H_2O flowing from an area of a lower solute to a higher solute concentration.

If a membrane is impermeable to a solute, H_2O flows across the membrane (i.e., osmosis) until the differences in solute concentrations have equilibrated.

Capillaries are impermeable to albumin (i.e., solute).

Increasing albumin concentration in the arteries and capillaries increases the movement of H_2O from the interstitial fluid to reduce the osmotic pressure in the arteries and capillaries.

45. B is correct.

Retinoblastoma protein (i.e., pRb or *RB1*) is a dysfunctional tumor suppressor protein.

pRb inhibits excessive cell growth by regulating the cell cycle through G1 (first gap phase) into S (DNA synthesis) phase.

pRb recruits chromatin-remodeling enzymes, such as methylases and acetylases.

46. E is correct.

Glycolysis occurs in the *cytoplasm.*

Krebs (TCA) cycle and *pyruvate oxidation* to acetyl-CoA occurs in the mitochondrial *matrix.*

Electron transport chain (ETC) uses cytochromes in the inner mitochondrial membrane and a proton (H^+) gradient in the intermembrane space).

A: *pyruvate* is oxidized to acetyl-CoA and transported from the cytoplasm into the matrix before joining oxaloacetate in the Krebs cycle.

B: Krebs cycle is the second stage of cellular respiration and occurs in the matrix of the mitochondrion.

C: *electron transport chain* is the final stage of cellular respiration occurring in the inner membrane (i.e., cytochromes) / intermembrane space (i.e., H^+ proton gradient) of the mitochondrion.

D: *reduction* (i.e., a gain of electrons) of FADH into $FADH_2$ occurs during the Krebs cycle.

47. D is correct.

Triglycerides are derived from glycerol and three fatty acids, commonly called "fats" and lipids, and are the most abundant lipid.

Teichoic acids are bacterial copolymers of carbohydrates and phosphate.

Peptidoglycan is formed by monosaccharide and amino acid polymers in bacterial cell walls.

Glycogen is a polysaccharide of glucose, a monosaccharide.

48. E is correct.

Secretory sequence in the flow of newly synthesized protein for export from the cell is:

rough ER → Golgi → plasma membrane

49. B is correct.

Minerals (e.g., potassium, sodium, calcium, and magnesium) are essential nutrients because they must be consumed in the diet and function as cofactors (i.e., nonorganic components) for enzymes.

Lysosomes do not digest minerals.

Organic molecules, such as nucleotides, proteins, and lipids, are hydrolyzed (i.e., degraded) into monomers.

Nucleotides have phosphate, sugar, and base; proteins have amino acids; lipids have glycerol and fatty acids.

50. D is correct.

Recycling of organelles within the cell is accomplished through autophagy by *lysosomes.*

51. B is correct.

Albumins are globular proteins in the circulatory system.

Carotenoids are organic pigments in chloroplasts of plants and photosynthetic organisms, such as some bacteria and fungi.

Carotenoids are fatty acid-like carbon chains containing *conjugated double bonds* and sometimes have six-membered carbon rings at ends.

Under visible light, they produce red, yellow, orange, and brown colors in plants and animals as pigments.

Waxes (esters of fatty acids and alcohols) are protective coatings on the skin, fur, leaves of higher plants, and on the exoskeleton cuticle of many insects.

Steroids (e.g., cholesterol, estrogen) have three fused cyclohexane rings and one cyclopentane ring.

Lecithin is an example of a phospholipid that contains glycerol, two fatty acids, a phosphate group, and nitrogen-containing alcohol.

52. A is correct.

Defective attachment of a chromosome to the spindle blocks activation of the *anaphase-promoting complex.*

Spindle checkpoint prevents anaphase onset in mitosis and meiosis until all chromosomes are attached to the spindle with the proper bipolar orientation.

53. B is correct.

Human gametes, formed during meiosis, are cells with a single copy (1N) of the genome.

After the second meiotic division, cells have a single unreplicated copy (i.e., devoid of a sister chromatid).

54. D is correct.

Smooth endoplasmic reticulum is the organelle for fatty acid, phospholipid, and steroid synthesis.

Smooth endoplasmic reticulum (SER) synthesizes lipids, phospholipids, and steroids; metabolism of carbohydrates and steroids; detoxification of alcohol and drugs; regulates Ca^{2+} concentration in muscle cells.

Endosome is a transport pathway starting at the Golgi.

Peroxisome undergoes beta-oxidation of long-chain fatty acids, which eventually yields $CO_2 + H_2O$.

Rough endoplasmic reticulum (RER) assembles proteins.

55. C is correct.

Centrioles are cylindrical structures mainly of *tubulin* in eukaryotic cells (except flowering plants and fungi).

Centrioles participate in the organization of the *mitotic spindle* and the completion of *cytokinesis.* They contribute to the structure of centrosomes and organize microtubules in the cytoplasm.

Centriole position determines the location of the nucleus and is crucial in the spatial arrangement of cells.

continued…

Plastid (e.g., chloroplast and chromoplast) are major organelles in plants and algae.

Plastids are the site of manufacturing and storing critical chemical compounds used by cells.

They often contain pigments used in photosynthesis. Pigments change to determine cell color.

Plastids, like prokaryotes, contain a circular double-stranded DNA molecule.

56. E is correct.

Centrioles are the organizational sites for microtubules (i.e., spindle fibers) that assemble during cell division (e.g., mitosis and meiosis).

Four phases of mitosis are *prophase*, *metaphase*, *anaphase*, and *telophase*.

Mitotic phases are followed by cytokinesis, which physically divides cells into two identical daughter cells.

Condensed chromosomes align along the *equatorial plane* in metaphase before the centromere (i.e., heterochromatin region of DNA) splits.

The two sister chromosomes begin their journey to the respective poles of the cell.

57. A is correct.

cAMP is a second messenger triggered when a ligand (e.g., peptide hormone or neurotransmitter) binds a membrane-bound receptor.

Adenylate cyclase (enzyme) is activated through a G-protein intermediate and converts ATP into cAMP.

Adenylate cyclase is attached to the inner layer of the phospholipid bilayer and is not in the cytoplasm.

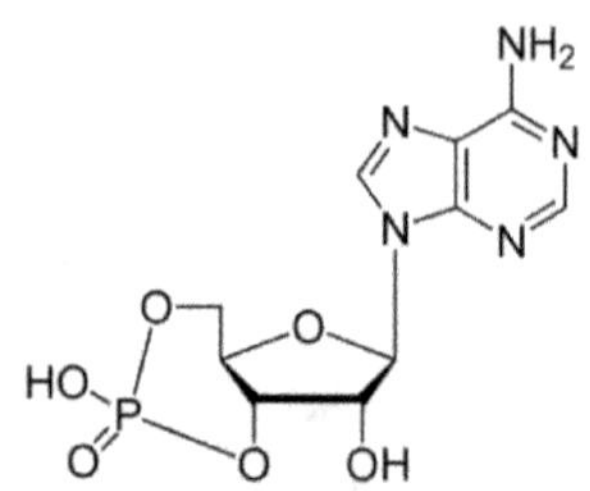

cAMP

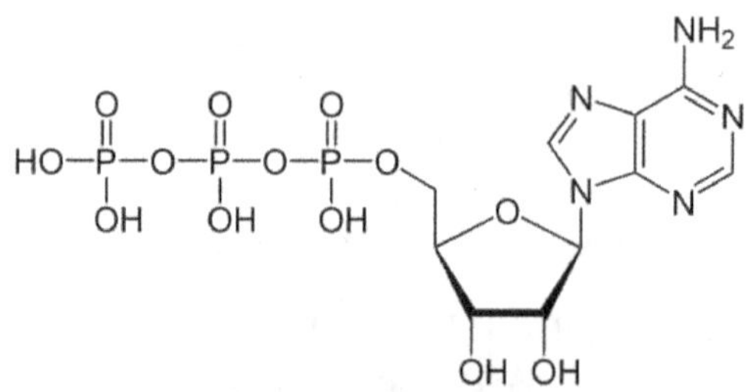

ATP

58. C is correct.

Cyclin and its associated *protein kinase* ensure a proper progression of cell division.

Cyclins are phosphorylated proteins responsible for specific events during cycle division (e.g., microtubule formation, chromatin remodeling).

Cyclins divide into *four classes* based on their behavior in the cell cycle: G1/S, S, M, and G1 cyclins.

59. A is correct.

Hypertonic solution is when there is a higher concentration of solutes outside the cell than inside.

When a cell is in a *hypertonic solution*, the tendency is for *water to flow out* of the cell to balance the concentration of the solutes.

Osmotic pressure draws water out of the cell, and the cell shrivels (i.e., crenation).

60. E is correct.

Peroxisomes are organelles abundant in the liver containing oxidases and detoxifying substances (i.e., alcohol and hydrogen peroxide).

61. B is correct.

Microtubules are hollow proteins composed of *tubulin* monomers necessary for:

1) formation of the *spindle apparatus* that separates chromosomes during cell division,

2) synthesis of *cilia* and *flagella*, and

3) formation of the *cytoskeleton* (within the cytoplasm).

II: actin and myosin are contractile fibers in muscle cells composed of microfilaments (not microtubules).

62. D is correct.

If a cell (e.g., pancreatic exocrine cell) is producing substantial amounts of proteins (i.e., enzymes) for export, this involves the *rough endoplasmic reticulum* (RER).

Protein synthesis begins in the nucleus with mRNA transcription, translated into polypeptides with the RER.

Exported proteins are packaged and modified in the Golgi; this cell would have a prominent nucleolus for rRNA (i.e., ribosomal components) synthesis.

63. A is correct.

Mitosis is cell division when new somatic (i.e., body) cells are added to multicellular organisms as they grow, and tissues are repaired or replaced.

Mitosis does not produce genetic variations. A daughter cell is identical in chromosome number and genetic makeup to the parental cell.

Eukaryotes divide by mitosis. Mitosis distributes identical genetic material to two daughter cells.

Fidelity of DNA transmission between generations without dilution is remarkable.

64. E is correct.

Amoeboids move using pseudopodia as bulges of cytoplasm powered by the elongation of flexible microfilaments (not microtubules).

Microtubules are long, hollow cylinders of polymerized α- and β-tubulin dimers critical in cellular processes such as maintaining cell structure and forming the cytoskeleton with microfilaments and intermediate filaments.

Microtubules comprise the internal structure of *cilia* and *flagella.*

Microtubules provide platforms for intracellular transport and participate in cellular processes, including the movement of secretory vesicles, organelles, and intracellular substances.

Microtubules engage in cell division (i.e., mitosis and meiosis), including the formation of mitotic spindles that pull apart eukaryotic chromosomes.

65. B is correct.

Apoptosis is programmed cell death that occurs during fetal development and aging.

Synaptic cleft development, the formation of separate digits in the hand of a fetus, and tadpole tail reabsorption are examples of apoptosis during development.

Synthesis of the uterine lining is an *anabolic* process that involves mitosis (i.e., cell division).

66. A is correct.

A cell involved in active transport (e.g., intestinal epithelial cells) requires much ATP.

Therefore, many mitochondria are needed to meet cellular respiration needs (i.e., glucose → ATP).

B: elevated levels of *DNA synthesis* occur in cells using mitosis for rapid reproduction (i.e., skin cells).

C: elevated levels of *adenylate cyclase* (i.e., cAMP second messenger) are in the target cells of peptide hormones.

D: *polyribosomes* are in cells with a high protein synthesis level.

E: *lysosomes* are organelles with low pH that function to digest intracellular molecules.

67. E is correct.

Coat proteins, like clathrin, form small vesicles to transport molecules within and between cells.

Endocytosis and *exocytosis* of vesicles allow cells to transfer nutrients, import signaling receptors, mediate immune responses, and degrade cell debris after tissue inflammation.

Endocytosis is when cells internalize receptor-ligand complexes from the cell surface (e.g., cholesterol bound to its receptor in the clathrin-coated pits).

Receptor-ligand complexes cluster in clathrin-coated pits at the cell surface and pinch off the vesicles that join acidic vesicles (i.e., endosomes).

68. D is correct.

Cells respond to peptide hormones through biochemical reactions involving membrane receptors and kinase activation.

First and second messengers communicate responses between cells because the G protein links the first and second messengers.

cAMP is a second messenger triggered when a ligand (e.g., peptide hormone or neurotransmitter) binds a membrane-bound receptor.

Adenylate cyclase (enzyme) is activated through a G-protein intermediate and converts ATP into cAMP.

Adenylate cyclase is attached to the inner layer of the phospholipid bilayer and is not in the cytoplasm.

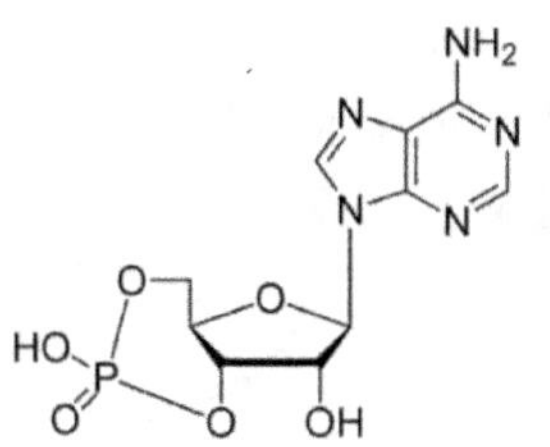

cAMP

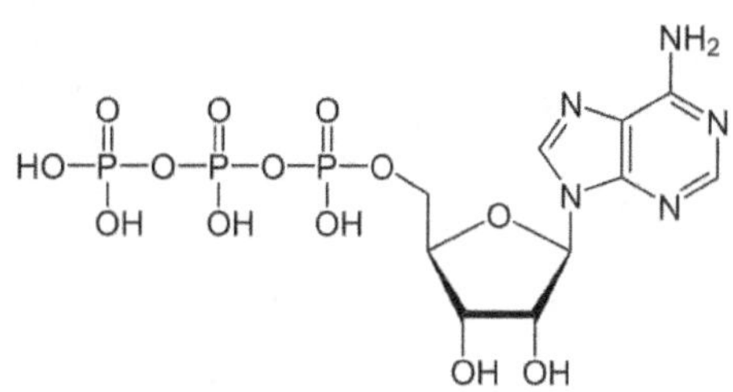

ATP

69. E is correct.

Transduction involves a virus and is one of three methods (*conjugation* and *transformation*) that bacterial cells use to introduce genetic variability into their genomes.

Transduction is when a virus introduces novel genetic material while infecting its host.

Mitosis occurs in somatic cells for growth and repair, producing two identical diploid (2N) daughter cells.

Notes for active learning

Notes for active learning

Notes for active learning

Cellular Metabolism and Enzymes – Detailed Explanations

1. D is correct.

α-helix (alpha helix) is a typical secondary structure of proteins.

α-helix is a *right-handed coiled* or spiral conformation (helix) when each backbone amino (N-H) group donates partial positive hydrogen to form a hydrogen bond with a lone pair of electrons from the backbone carbonyl (C=O) group of another amino acid four residues earlier.

2. A is correct.

ATP is a nucleotide composed of adenosine (A), ribose, and three phosphate groups.

3. C is correct.

Hydrophobic side chain groups (e.g., phenylalanine, methionine, leucine, and valine) interact through hydrophobic interactions, excluding water from the attraction region.

Nine amino acids with hydrophobic side chains are alanine (Ala), glycine (Gly), isoleucine (Ile), leucine (Leu), methionine (Met), phenylalanine (Phe), proline (Pro), tryptophan (Trp), and valine (Val).

B: *hydrogen bonds* require hydrogen to be connected to an electronegative atom of fluorine, nitrogen, oxygen, or chlorine (weaker hydrogen bond due to the larger valence shell size) to establish a $\delta+$ and $\delta-$ (partial positive and negative regions that attract).

D: *cysteine* is the only amino acid capable of forming disulfide bonds because the side chain contains sulfur.

E: *dipole-dipole interactions* (collectively Van der Waals) are weak interactions from the close association of permanent or induced dipoles.

Proteins contain many dipole interactions, which vary considerably in strength.

4. B is correct.

Cellular respiration uses enzyme-catalyzed reactions as a catabolic pathway using potential energy of glucose.

While glycolysis yields two ATP per glucose, cellular respiration (i.e., glycolysis, Krebs/TCA cycle, and electron transport chain) produces about 30–32 ATP.

Around 2004, research reduced the estimate of 36-38 ATP due to inefficiencies of ATP synthase.

Cellular respiration is an aerobic process whereby O_2 is the final acceptor of electrons passed from cytochrome electron carriers during the final stage of the electron transport chain.

Glycolysis (glucose to pyruvate) occurs in the cytoplasm.

Krebs cycle (TCA) occurs in the matrix (i.e., the cytoplasm of mitochondria), while the electron transport chain occurs in the intermembrane space of the mitochondria for eukaryotes.

For prokaryotes, both processes occur in the cytoplasm because prokaryotes lack mitochondria.

continued...

Some forms of anaerobic respiration use an electron acceptor other than O_2 (e.g., iron, cobalt, or manganese reduction) at the end of the electron transport chain.

Glucose molecules are entirely oxidized in these cases because glucose products enter the Krebs cycle.

Highly reduced chemical compounds (e.g., NADH or $FADH_2$) establish an electrochemical gradient across a membrane in the Krebs cycle.

The reduced chemical compounds (NADH or $FADH_2$) are oxidized by integral membrane proteins that transfer the electron to the final electron acceptor.

5. E is correct.

K_m is the substrate concentration [S] at half the maximum reaction velocity (V_{max}) and increases when the antibody binds the substrate.

This reaction kinetics follow the profile of a *competitive inhibitor* (visualized by the y-axis intercept) in the Lineweaver-Burke (double reciprocal).

The substrate binds the antibody and is not available to react with the enzyme; therefore, an additional substrate is needed to bind the same amount of enzyme (as a lower substrate concentration in the absence of an antibody).

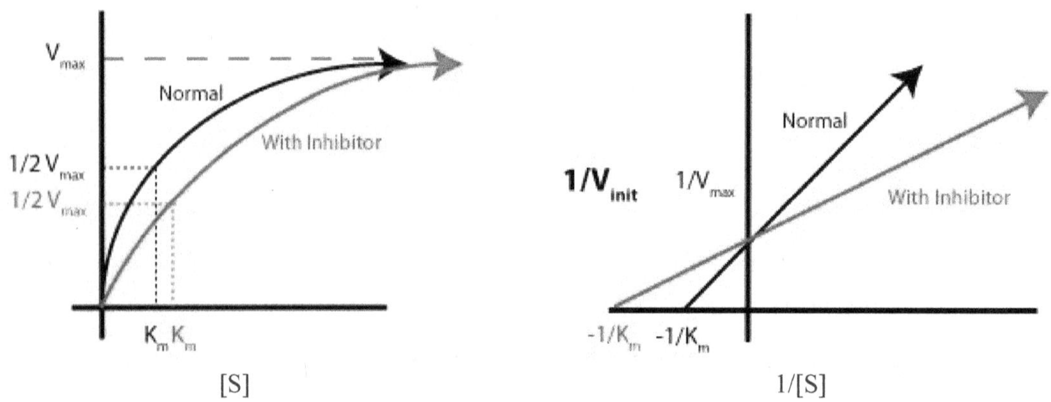

Michaelis-Menten equation (left) and Lineweaver-Burke (double reciprocal) plot

Graph on *left* illustrates the Michaelis-Menten equation for enzyme kinetics.

Graph on *right* is the Lineweaver-Burke (double reciprocal) plot showing the change in K_m when the substrate is subjected to competitive inhibition.

V_{max} does not change when the number of available substrates changes.

Binding to a substrate does not affect V_{max} but affects K_m because the amount of available substrate is reduced, altering the reaction kinetics because it binds the substrate.

6. C is correct.

Metabolism means change and refers to life-sustaining chemical transformations within cells.

Enzyme-catalyzed reactions allow organisms to maintain structures, grow, reproduce, and respond.

Metabolism includes *anabolism* (or *buildup*) and *catabolism* (or *breakdown*).

Catabolism is the breakdown of organic matter and harvests energy from cellular respiration (e.g., glycolysis, Krebs cycle, and the electron transport chain).

Anabolism utilizes energy (e.g., ATP) to *synthesize biomolecules* (e.g., lipids, nucleic acids, proteins).

7. D is correct.

Fermentation is an anaerobic process and occurs in the absence of oxygen.

Purpose of fermentation is to *regenerate* the high-energy nucleotide intermediate of NAD^+.

Oxidative phosphorylation occurs during the electron transport chain as the final stage of cellular respiration.

8. A is correct.

Acid hydrolysis has several effects:

1) partial destruction of tryptophan prevents the proper estimate of the tryptophan concentration;

2) conversion of asparagine into aspartic acid prevents the direct measure of asparagine;

3) conversion of glutamine to glutamic acid.

Therefore, the concentration of glutamic acid is an indirect measure of glutamine.

9. B is correct.

When plasma glucose levels are low, the body utilizes other energy sources for cellular metabolism.

Energy sources are used in preferential order:

glucose → other carbohydrates → fats → proteins

These molecules are converted to glucose or glucose intermediates degraded in the glycolytic pathway and the Krebs cycle (i.e., citric acid cycle).

Proteins are used last for energy because there is no protein storage in the body.

Catabolism of protein results in muscle wasting and connective tissue breakdown, which is harmful over time.

10. E is correct.

Activation energy is the minimum amount of energy needed for a reaction. It is considered an "energy barrier" because energy must be added to the system.

Enzymes act by *decreasing the activation energy* (or *lowering the energy barrier*), allowing the reaction, and increasing the reaction rate.

11. A is correct.

Phosphate group from ATP is transferred to glucose, creating a *phosphodiester bond.*

ATP + Glucose → Glucose-6-phosphate + ADP

12. D is correct.

Second messengers are common in eukaryotes.

Cyclic nucleotides (e.g., cAMP and cGMP) are second messengers transmitting extracellular signals from the cell membrane to intracellular proteins.

Second messengers are used by hydrophilic protein hormones (e.g., insulin, adrenaline) that cannot cross the plasma membrane.

Steroid hormones (e.g., testosterone, estrogen, progesterone) are lipid-soluble and can pass through the plasma membrane and enter the cytoplasm without second messengers.

13. E is correct.

Greatest direct source of ATP synthesis involves the *electron transport chain.*

Glycolysis occurs in the cytoplasm, the *Krebs cycle* (TCA) in the matrix of mitochondria, and *oxidative phosphorylation* (i.e., the electron transport chain) in the intermembrane space and inner membrane of mitochondria.

Glycolysis produces pyruvate, converted into acetyl CoA, and joined to oxaloacetate in the Krebs cycle, producing citrate as the first intermediate.

Krebs cycle produces two GTP (i.e., ATP) per glucose molecule (or 1 GTP per pyruvate).

Most ATP is formed by *oxidative phosphorylation* when NADH and $FADH_2$ nucleotides are oxidized and donate their electrons to the cytochromes in the electron transport chain.

14. D is correct.

V_{max} is the reaction velocity at a fixed enzyme concentration and depends on the total enzyme concentration. Adding enzymes allows more enzyme reactions per minute.

II: V, not V_{max}, depends on [S]. V_{max} is a constant for a specified amount of enzyme.

III: *competitive inhibition* is when the inhibitor binds reversibly to the active site. Adding enough substrate overcomes competitive inhibition, and the original V_{max} is obtained.

15. B is correct.

Two major models of *enzyme-substrate binding*:

Induced-fit model is when the initial interactions between enzyme and substrate (e.g., hexokinase and glucose) are weak, but the weak interactions induce *conformational changes* that strengthen binding.

Lock-and-key model is when the molecules conform for binding and do not change.

16. D is correct.

Gibbs free energy (ΔG) of a reaction is not dependent on the amount of enzyme.

Gibbs free energy measures the energy difference (Δ) between the reactants and products.

> ΔG is *positive* for endergonic reactions with products higher in energy than the reactants (absorbs energy).
>
> ΔG is *negative* for exergonic reactions when products are lower in energy than the reactants (releases energy).

17. C is correct.

Zymogens are inactive forms of enzymes and have an ~ogen suffix (e.g., pepsinogen).

From the question, glucokinase has a higher Michaelis constant (K_m).

A higher value for K_m is due to enzymes having a lower affinity for the reactant (e.g., glucose).

Information about K_m does not answer the question and is merely a distraction.

A: *hexokinase* is a control point enzyme (i.e., irreversible steps) regulated by negative feedback inhibition.

B: *hexokinase* and glucokinase catalyze the same reaction (from the question); by definition, *isozymes.*

D: *fructose* is not a reactant in glycolysis.

18. B is correct.

Secondary structure (2°) is the repetition of α-helices or β-pleated sheets in the polypeptide backbone.

Primary structure (1°) is the linear sequence of amino acids.

Tertiary structure (3°) involves interactions between the side chains of the amino acids.

Quaternary structure (4°) requires the interaction of two or more polypeptides. 4° requires more than one polypeptide chain in the mature protein (e.g., hemoglobin).

19. E is correct.

Cofactors are organic or inorganic molecules classified depending on how tightly they bind to an enzyme.

Coenzymes are loosely bound *organic cofactors* released from the enzyme's active site during the reaction (e.g., ATP, NADH).

Prosthetic groups are tightly bound organic cofactors.

20. B is correct.

Quaternary (4°) *structure* of proteins involves two or more polypeptide chains.

Hydrophobic interactions and disulfide bridges between cysteines maintain the quaternary structure between the different polypeptide chains (e.g., 4 chains of 2 α and 2 β in hemoglobin).

21. A is correct.

Mitochondrion is the organelle bound by a double membrane, where the Krebs (TCA) cycle, electron transport, and oxidative phosphorylation occur.

In eukaryotes, the Krebs cycle (i.e., TCA) occurs in the matrix of the mitochondria.

Matrix is an interior space in the mitochondria analogous to the cell cytoplasm.

B: *smooth ER* (endoplasmic reticulum) is connected to the nuclear envelope and synthesizes lipids, phospholipids, and steroids.

Cells secreting lipids, phospholipids, and steroids (e.g., testes, ovaries, skin oil glands) have prominent smooth endoplasmic reticulum.

Smooth ER carries out the metabolism of carbohydrates, drug detoxification, attachment of receptors on cell membrane proteins, and steroid metabolism.

C: *cytosol* is common to eukaryotes and prokaryotes and equivalent to the cytoplasm.

D: *nucleolus* is the organelle within the nucleus responsible for synthesizing ribosomal RNA (rRNA).

E: *intermembrane space of the mitochondria* is where the electron transport chain (ETC) establishes a proton gradient (or chemiosmotic gradient) with concentrated H^+ ions forced into the space in eukaryotes.

22. B is correct.

Pyruvate is the product of glycolysis, converted into acetyl Co-A as the starting reactant for the Krebs cycle.

Pyruvate is *not* a waste product of cellular respiration. Metabolic waste products are from enzymatic processes.

Pyruvate, made during glycolysis (as an intermediate of cellular respiration), is converted to acetyl CoA and enters the Krebs cycle to be oxidized into CO_2 and H_2O or converted into the *waste product of lactate* under anaerobic conditions. Waste products can damage cells and must be removed.

A: *lactate* (2 carbons) is produced from *pyruvate* (3-carbon chain) as the waste product of anaerobic respiration.

Lactate is converted to pyruvate in the liver when O_2 becomes available.

If lactate is not metabolized, it can lead to lactic acidosis (acidification of the blood) and death.

C and D: CO_2 and H_2O are waste products of aerobic respiration.

CO_2 and H_2O are removed from the lungs via expiration during regular breathing.

CO_2 increases lead to acidosis (lowering blood pH) because CO_2 reacts with H_2O to form carbonic acid (H_2CO_3).

E: a*mmonia* is the waste product of protein metabolism, converted to the less toxic urea in the liver and removed by the kidneys.

If ammonia is not converted and cleared, the blood becomes alkaline (higher in pH), potentially fatal.

23. A is correct.

V_{max} is proportional to the number of active sites on the enzyme (i.e., [enzyme] or enzyme concentration).

K_m remains constant because it is a measure of the active site affinity for the substrate.

Regardless of the number of enzyme molecules, each enzyme interacts with the substrate similarly.

Models propose mechanisms of enzymatic catalysis.

Michaelis–Menten model describes enzyme-substrate interactions, whereby enzymatic catalysis occurs at a specific site on enzymes – the active site.

Substrate (S) binds the enzyme (E) at the active site for an enzyme-substrate (ES) complex.

ES complex dissociates into enzyme and substrate or moves forward in the reaction to form product (P) and original enzyme.

$$E + S \underset{k_{-1}}{\overset{k_1}{\rightleftharpoons}} ES \xrightarrow{k_2} P + E$$

Rate constants for different steps in the reaction are k_1, k_{-1}, and k_2.

The overall rate of product formation is a combination of each rate constant.

At constant enzyme concentrations, varying [S] changes the rate of product formation.

At low [S], a small fraction of enzyme molecules is occupied with the substrate.

ES and [P] formation rate increases linearly with substrate concentration.

At high [S], all active sites are occupied with the substrate, and increasing the substrate concentration further does not increase the reaction rate, and the reaction rate is V_{max}.

Relationship between substrate concentration and reaction rate:

$$V = \frac{[S]}{[S] + K_m} V_{max}$$

24. E is correct.

Apoenzyme, with its cofactors, is a *holoenzyme* (an active form).

Enzymes that require a cofactor, but do not have one bound, are *apoenzymes* (or apoproteins).

Cofactors can be inorganic (e.g., metal ions) or organic (e.g., vitamins).

Organic cofactors are coenzymes or prosthetic groups.

Most cofactors are not covalently attached to an enzyme but are tightly bound.

Organic prosthetic groups can be covalently bound.

Holoenzyme refers to enzymes containing multiple protein subunits (e.g., DNA polymerases), whereby the holoenzyme is the complete complex with all subunits needed for the activity.

25. C is correct.

ATP releases about –7.4 kcal/moles.

As an approximation, 1 kcal/mol = ~ 4.2 kJ/mol

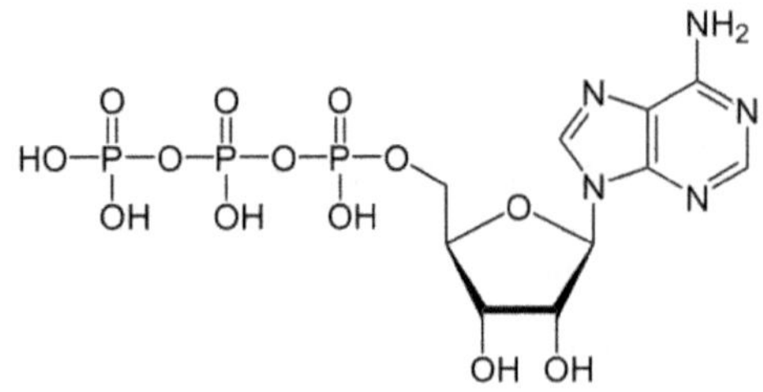

Structure of the nucleotide of ATP
(3 phosphates + ribose sugar + adenosine)

ATP contains three phosphate groups and is produced by enzymes (e.g., ATP synthase) from adenosine diphosphate (ADP) or adenosine monophosphate (AMP) and phosphate group donors.

Three central mechanisms of ATP biosynthesis are *oxidative phosphorylation* (in cellular respiration), *substrate-level phosphorylation*, and *photophosphorylation* (in photosynthesis).

Metabolic processes using ATP convert it back into its precursors; ATP is continuously recycled.

26. D is correct.

Cysteine is the only amino acid capable of forming a *disulfide* (S–S covalent) bond and serves an essential structural role in many proteins.

Thiol side chain in cysteine often participates in enzymatic reactions, serving as a nucleophile.

Thiol is susceptible to oxidization to give the *disulfide* derivative *cystine.*

27. C is correct.

Electron transport chain (ETC) uses cytochromes in the inner mitochondrial membrane and is a complex carrier mechanism of electrons that produces ATP through *oxidative phosphorylation.*

A: Krebs (i.e., TCA) cycle occurs in the *matrix* of the mitochondria.

Krebs cycle begins when acetyl CoA (i.e., 2-carbon chain) combines with oxaloacetic acid (i.e., 4-carbon chain) to form citrate (i.e., 6-carbon chain).

A series of additional enzyme-catalyzed reactions result in the oxidation of citrate and the release of two CO_2, and oxaloacetic acid is regenerated to begin the cycle again once it is joined to an incoming acetyl CoA.

B: *long-chain fatty acid degradation* occurs in peroxisomes that hydrolyze fat into smaller molecules fed into the Krebs cycle and used for cellular energy.

D: *glycolysis* occurs in the cytoplasm as the oxidative breakdown of glucose (i.e., 6-carbon chain) into two pyruvates (i.e., 3-carbon chain).

E: *ATP synthesis* occurs in the cytoplasm (glycolysis), the mitochondrial matrix (Krebs cycle), and the inner mitochondrial membrane (ETC).

28. E is correct.

Apoenzyme is an inactive enzyme without its cofactor.

Apoenzyme, with its cofactor(s), is a *holoenzyme* (an active form).

Enzymes that require a cofactor, but do not have one bound, are *apoenzymes* (or apoproteins).

Cofactors are inorganic (e.g., metal ions) or organic (e.g., vitamins).

Organic cofactors are *coenzymes* or *prosthetic groups*.

29. B is correct.

Disulfide bond is a covalent bond derived by coupling two thiols (~S-H) groups as R–S–S–R.

Peptide bonds join adjacent amino acids and make the protein's linear sequence (i.e., primary structure).

30. B is correct.

Enzyme prosthetic groups attach via strong molecular forces (e.g., covalent bonds).

A: *Van der Waals* interactions, like dipole/dipole, are weak molecular forces.

C: *ionic bonds* are powerful in non-aqueous solutions but are not used to attach prosthetic groups because they are weak in aqueous environments such as plasma. For example, NaCl dissociates in water.

D: *hydrogen bonds* (like hydrophobic interactions) are an example of a weak molecular force.

E: *dipole-dipole interactions* are an intermolecular force intermediate in strength between hydrogen bonds and van der Waals interactions that require a dipole (separation in charge) within the bonding molecules.

31. E is correct.

Cellular respiration begins via glycolysis in the cytoplasm but is completed in the mitochondria.

Glucose (six-carbon chain) is cleaved into two pyruvate molecules (three-carbon chain).

Pyruvate is converted to acetyl CoA that enters the Krebs cycle in the mitochondria.

NADH (from glycolysis and the Krebs cycle) and $FADH_2$ (from the Krebs cycle) enter the electron transport chain on the inner membrane of the mitochondria.

32. C is correct.

Primary activity of the thyroid hormone is to *increase* the *basal metabolic rate*.

33. D is correct.

Cofactors are organic or inorganic molecules classified depending on how tightly they bind to an enzyme.

Coenzymes are loosely bound organic cofactors released from the enzyme's active site during the reaction (e.g., ATP, NADH).

Prosthetic groups are tightly bound organic cofactors.

34. B is correct.

Primary structure is the *linear sequence* of amino acids in the polypeptide.

35. D is correct.

Denaturing the polypeptide disrupts the 4°, 3° and 2° structures, while the 1° structure (i.e., linear sequence of amino acids) is unchanged.

In folded proteins, 1° structure determines subsequent 2°, 3° and 4° structures.

36. B is correct.

$FADH_2$ is produced in the Krebs cycle (TCA) in cellular respiration.

A: product of glycolysis is pyruvate (i.e., pyruvic acid), which is converted to *lactic acid* (in humans) or *ethanol* & *CO_2* (in yeast) *via* anaerobic conditions.

Yeast is used for alcoholic production and baking for the dough to rise.

C: in glycolysis, the first series of aerobic (or anaerobic) reactions occur in the cytoplasm by breaking the 6-carbon glucose into two 3-carbon pyruvate molecules.

D: two ATP are required, and four ATP are produced = net of two ATP at the end of glycolysis.

E: two NADH are produced by glycolysis and are oxidized (i.e., lose electrons) to compound Q in the electron transport chain within the mitochondria.

37. A is correct.

There are 13 essential (i.e., must be consumed) vitamins: A, B_1 (thiamine), B_2 (riboflavin), B_3 (niacin), B_5 (pantothenic acid), B_6 (pyridoxine), B_7 (biotin), B_9 (folate) and B_{12} (cobalamin), C, D, E and K.

Four fat-soluble vitamins, A, D, E, and K are stored in adipose tissue.

38. D is correct.

Hemoglobin is a *quaternary protein* with two α and two β chains with Fe^{2+} (or Fe^{3+}) as the prosthetic group.

Prosthetic groups are *tightly bound* cofactors (i.e., nonorganic molecules such as metals).

39. E is correct.

Glycine is the only achiral amino acid because hydrogen is the R group (i.e., side chain).

Optical activity is the ability of a molecule to rotate plane-polarized light and requires a chiral center (i.e., carbon atom bound to 4 different substituents).

D: *cysteine* contains sulfur in the side chain and is the only amino acid forming disulfide bonds (i.e., S–S).

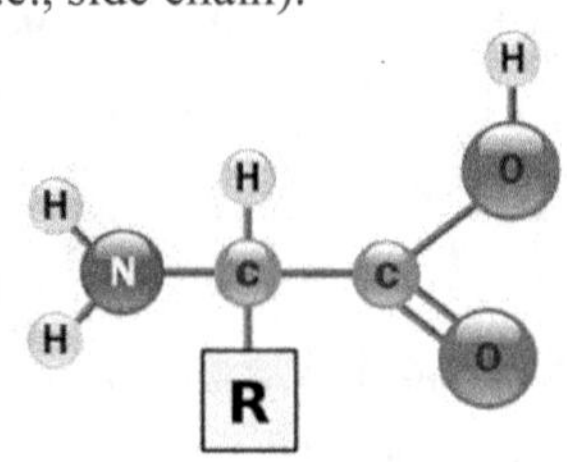

Amino acid structure with the α-carbon attached to the side chain R

40. A is correct.

During glycolysis, two net (four gross = total) molecules of ATP are produced by *substrate-level phosphorylation*, and two molecules of NAD^+ are reduced (i.e., gain electrons) to form NADH.

B: during glycolysis, *two pyruvates* form from each starting molecule of glucose.

C: *glycolysis* is an anaerobic process and does not require O_2.

D: *glucose* is partially *oxidized* (not reduced) into *two pyruvate* molecules during glycolysis.

E: *pyruvate* is produced during glycolysis (not in the Krebs cycle), converted to acetyl CoA (coenzyme A) that enters the Krebs cycle.

41. B is correct.

Krebs (TCA) cycle and the *electron transport chain* are metabolic processes in the mitochondria.

42. C is correct.

Feedback inhibition occurs when a product binds an enzyme to prevent it from catalyzing further reactions.

Since it is *allosteric inhibition*, this product binds a site other than the active site and changes the conformation of the active site to inhibit the enzyme.

Although the allosteric inhibitor can be the same product of the enzyme-catalyzed reaction *or* the product of the metabolic pathway, it is usually the final product that inhibits the entire metabolic pathway, thus saving the energy necessary to produce the intermediates as well as the final product.

When levels of the final product are low, the enzyme resumes its catalytic activity.

43. A is correct.

Obligate anaerobe reactions occur in the *absence* of O_2.

Oxidative phosphorylation occurs during the electron transport chain and requires a molecule of *oxygen* as the final acceptor of electrons shuttled between the cytochromes.

44. E is correct.

Peptide bonds join adjacent amino acids.

The amino acid (written on the left by convention) contributes an amino group ($\sim NH_2$).

The other amino acid (written on the right by convention) contributes a carboxyl group (~COOH) that undergoes condensation (i.e., joining of two pieces to form a connected unit) *via* dehydration (i.e., removal of H_2O during bond formation).

45. A is correct.

Oxidative phosphorylation (O_2 as substrate/reactant) occurs only in the electron transport chain (the last step for cellular respiration) in the *inner membrane* (i.e., cristae) of the mitochondria.

NADH and $FADH_2$ donate electrons to a series of cytochrome (i.e., protein) molecules embedded in the inner membrane and create an electron gradient, establishing a proton (H^+) gradient within the *intermembrane space* of the mitochondria.

H^+ gradient drives a proton pump coupled with an enzyme, producing ATP via *oxidative phosphorylation.*

Electrons from NADH and $FADH_2$ are transferred to ½ O_2 (i.e., oxidative phosphorylation) to generate ATP and form H_2O as a metabolic waste (along with CO_2) from cellular respiration.

H_2O and CO_2 (as metabolic wastes) are expired from the lungs during breathing.

Each Krebs cycle $FADH_2$ yields two ATP, while each NADH yields three ATP in the electron transport chain.

NADH from glycolysis only produces two ATP because energy is expended to shuttle the NADH from glycolysis (in the cytoplasm) through the double membrane of the mitochondria.

46. C is correct.

Enzymes do not affect free energy (ΔG); decreasing activation energy increases the reaction rate.

Enzymes do *not* affect the position or direction of equilibrium; they only affect the speed of equilibrium.

47. C is correct.

Zymogen (or proenzyme) is an inactive enzyme precursor. Proteolysis (i.e., cleavage) of the zymogen makes it active to catalyze reactions. Changes in substrate concentration regulate enzyme activity; an increase in substrate concentration increases enzyme activity (at saturation, V_{max} is achieved).

Post-translational modifications occur during protein biosynthesis (after translation), and they may involve cleaving proteins or introducing new functional groups to arrive at the mature protein product.

The enzyme may not be functional until post-translational modifications, a crucial regulatory aspect.

48. B is correct.

2° structure consists of numerous local conformations, such as α-helixes and β-sheets.

Secondary conformations are stabilized by hydrogen bonds or the covalent disulfide bonds between S–S of two cysteines to form cystine.

To be classified as 2° structure, the bonds must be formed between amino acids within about 12-15 amino acids along the polypeptide chain.

Otherwise, interactions (e.g., H bonds or disulfide bonds) that connect amino acid residues of more than 15 amino acids along the polypeptide chain qualify as 3° protein structure.

If the bonds occur between different polypeptide strands, the interactions qualify as 4° protein structures (e.g., two α and two β chains in hemoglobin).

49. E is correct.

Glycogen is a polysaccharide (i.e., carbohydrate) storage form with *highly branched glucose monomers.*

Extensive branching in glycogen provides numerous ends to the molecule to facilitate the rapid hydrolysis (i.e., cleavage and release) of individual glucose monomers when needed *via* epinephrine (i.e., adrenaline).

Glycogen is synthesized in the liver because of high plasma glucose concentrations and stored in muscle cells for release during exercise.

A: *glycogenesis* involves the synthesis of glycogen.

B: *glycogenolysis* involves the degradation of glycogen.

D: plants produce *starch* (i.e., analogous to glycogen) as their storage carbohydrate.

50. B is correct.

The fraction of occupied active sites for an enzyme equals V/V_{max} which = [S]/([S] + K_m).

If [S] = 2K_m, the fraction of occupied active sites is 2/3.

51. C is correct.

Allosteric regulators bind enzymes at allosteric sites (i.e., other than active sites) with *non-competitive inhibition.*

Once bound to the allosteric site, the enzyme changes its conformational shape.

If the shape change causes the ligand (i.e., molecule destined to bind to the active site) to bind to the active site less efficiently, the modulator is *inhibitory.*

Modulator is excitatory if the shape change causes the ligand to bind to the active site more efficiently (i.e., the active site becomes open and accessible).

52. B is correct.

Biological reactions can be *exergonic* (releasing energy) or *endergonic* (consuming energy).

Biological reactions have *activation energy*, the minimum amount required for the reaction.

These reactions use *enzymes* (i.e., *biological catalysts*) to lower activation energy and increase the reaction rate.

53. C is correct.

Oxidative phosphorylation (in the electron transport chain) uses ½ O_2 as the ultimate electron acceptor.

Electron transport chain (ETC) is on the inner mitochondrial membrane. It uses cytochromes (i.e., proteins) in the inner membranes of the mitochondria to pass electrons released from the oxidation of NADH & $FADH_2$.

Mitochondrial matrix (i.e., the cytoplasm of mitochondria) is the site for the Krebs cycle (TCA).

Outer mitochondrial membrane does not directly participate in oxidative phosphorylation or the Krebs cycle.

Glycolysis and Krebs cycle produce ATP via *substrate-level phosphorylation*, while *oxidative phosphorylation* requires O_2 and occurs during the ETC.

54. D is correct.

Enzymes are often proteins and function as biological catalysts at an optimal temperature (physiological temperature of 36 °C) and pH (7.35 is blood pH).

At higher temperatures, proteins *denature* (i.e., unfolding by disrupting hydrogen and hydrophobic bonds and changing shape) and lose their function.

Proteins often interact with inorganic minerals (i.e., *cofactors*) or organic molecules (i.e., *coenzymes or tightly bound prosthetic groups*) for optimal activity.

Mutations affect *DNA sequences* that encode proteins, resulting in a change in the amino acid sequence of the polypeptide and a change in conformation within the enzyme.

A change in the conformation of enzymes often results in a change in function.

55. D is correct.

Gibbs free energy is:

$$\Delta G = \Delta G° + RT\ln Q$$

where $\Delta G°$ = the Gibbs free energy change per mole of reaction for unmixed reactants and products at standard conditions (i.e., 298K, 100kPa, 1M of each reactant and product), R = the gas constant (8.31 $J \cdot mol^{-1} \cdot K^{-1}$), T = absolute temperature and Q = [product] / [reactant]

56. D is correct.

Allosteric enzymes have allosteric sites that bind metabolites.

Allosteric sites are distinct from the active sites that bind substrate.

An enzyme must have a *quaternary structure* (multiple polypeptides) to bind more than one molecule.

Allosteric enzymes show *cooperative* substrate binding, with the binding of a molecule at one site affecting the binding at another.

If an enzyme binds an inhibitor at an allosteric site, this *decreases the affinity* of the active site for the substrate.

57. C is correct.

Monoamine oxidases (MAO) are a family of enzymes that catalyze the oxidation of monoamine neurotransmitters. Monoamine neurotransmitters and neuromodulators contain one amino group connected to an aromatic ring by a two-carbon chain ($\sim CH_2–CH_2$).

Monoamines are derived from the aromatic amino acids phenylalanine, tyrosine, tryptophan, and thyroid hormones by aromatic amino acid decarboxylase (enzyme). Monoamines trigger crucial components such as emotion, arousal, and cognition.

Monoamine oxidases are linked to some psychiatric and neurological disorders.

High or low levels of MAOs are associated with schizophrenia, depression, attention deficit disorder, substance abuse, and migraines. MAO degrades serotonin, melatonin, norepinephrine, and epinephrine.

Excessive levels of catecholamines (e.g., epinephrine, norepinephrine, dopamine) leads to hypertensive crisis (e.g., hypertension), and excessive serotonin leads to serotonin syndrome (e.g., increased heart rate, perspiration, dilated pupils).

58. C is correct.

ΔG for hydrolysis of ATP to ADP and P*i* is –7.3 kcal/mole.

59. C is correct.

I: *prosthetic groups* are tightly bound chemical compounds required for the enzyme's catalytic activity, and it is often involved at the active site.

II: *zymogens* are inactive enzyme precursors requiring biochemical changes to become active enzymes.

This activation is usually a proteolysis (protein cleaving) reaction that reveals the active site.

60. B is correct.

Hydrolysis of ATP → ADP + phosphate forms glucose-6-phosphate from glucose and phosphate.

Enzymatic coupling of reactions allows ATP hydrolysis energy to drive the synthesis of glucose-6-phosphate.

61. A is correct.

Protein function can be altered by *post-translational modifications* (often in Golgi) after synthesis.

The common post-translational modification in eukaryotes is *phosphorylation* which forms the "*mature*" protein. Phosphorylation is the "activator" *or* "deactivator" of proteins/enzymes.

Secondary structure of proteins (e.g., alpha helix and beta pleated sheets) is stabilized by hydrogen bonds and hydrophobic interactions.

Quaternary structure requires two or more polypeptide chains.

Integral membrane proteins are embedded within the plasma membrane. Integral proteins are amphipathic (i.e., hydrophobic and hydrophilic portions). They have a similar structural motif with a hydrophobic domain, α-helical or β-sheet, which traverses the hydrophobic core of the lipid-bilayer membrane.

62. D is correct.

CO_2 is a product of alcohol fermentation (ethanol) and pyruvate oxidation (lactic acid) under aerobic conditions.

In yeast, *alcoholic fermentation* occurs in an anaerobic (i.e., without oxygen) process, producing ATP and the waste products ethanol (alcohol) and carbon dioxide (CO_2).

Pyruvate oxidation occurs in the mitochondria and cytosol in eukaryotes and prokaryotes, respectively.

Pyruvate dehydrogenase complex (PDC) has enzymes that oxidize pyruvate to acetyl-CoA, NADH, and carbon dioxide waste products.

Under anaerobic conditions (e.g., exercise in humans), pyruvate is converted to 2 moles of *lactic acid,* and NADH is oxidized (i.e., loss of hydrogen atom) to NAD^+.

63. E is correct.

Glucose and fructose have different chemical properties despite the same *molecular formula* ($C_6H_{12}O_6$).

Molecular formula of fructose and glucose are identical; however, the *structural formulas* (shown below) are different, giving rise to many *physical and chemical differences* between the two monosaccharides.

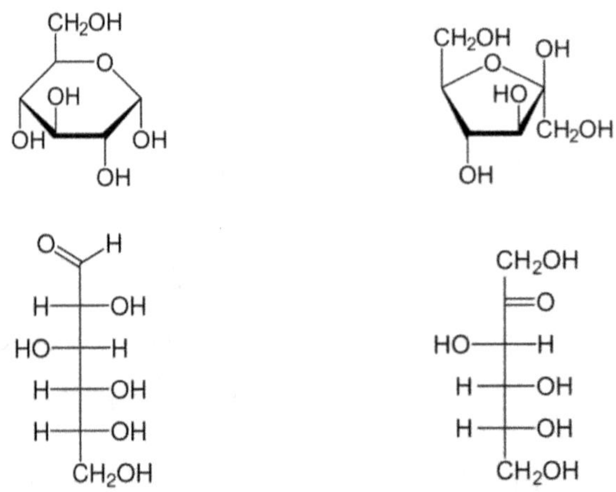

Haworth projections (above) and Fisher projections (below) for glucose on the left and fructose on right

Glucose has an aldehyde (i.e., terminal carbonyl) functional group; fructose is a ketone (i.e., internal carbonyl).

Compared to glucose, fructose has an increased density and decreased melting point with an increased BP.

Glucose, commonly referred to as blood sugar, is the "*preferred*" metabolic energy source.

Insulin is secreted from the *beta cells of the pancreas* in response to increased blood sugar (i.e., glucose) and facilitates glucose's entry into cells. Fructose does *not* stimulate the release of insulin as glucose does.

Glucose fuels cellular respiration, but fructose does not.

Glucose relies on glucokinase or hexokinase enzymes.

Fructose is more lipogenic and relies on fructokinase to initiate metabolism.

Fructose is the sweetest natural carbohydrate.

Notes for active learning

Notes for active learning

Photosynthesis – Detailed Explanations

1. E is correct.

C_4 plants use C_4 carbon fixation.

There are three carbon fixation mechanisms: C_3, C_4 and CAM (*crassulacean acid metabolism*).

C_4 is named the 4-carbon molecule in carbon fixation compared to a 3-carbon molecule product in C_3 plants.

C_4 plants have a competitive advantage over plants that use the C_3 carbon fixation pathway under drought, hot temperatures, and nitrogen or CO_2 limitation conditions.

About 8,100 plant species use C_4 carbon fixation, representing about 3% of terrestrial species of plants.

C_4 and CAM overcome the tendency of the RuBisCO to fix O_2 rather than CO_2 in photorespiration. They use a more efficient enzyme to fix CO_2 in mesophyll cells and shuttling this fixed carbon to bundle-sheath cells.

In bundle-sheath cells, RuBisCO is isolated from atmospheric oxygen and saturated with CO_2 released by decarboxylation of malate or oxaloacetate.

C_4 fixation is hypothesized to have evolved later.

Because these additional steps require more energy from ATP, C_4 plants can efficiently fix carbon in specific conditions, with the C_3 pathway being efficient in other conditions.

2. D is correct.

CAM (*crassulacean acid metabolism*) photosynthesis is a carbon fixation pathway that evolved in some plants to adapt to arid conditions.

In a plant using CAM, the stomata in the leaves remain shut during the day to reduce evapotranspiration but open at night to collect carbon dioxide (CO_2).

CO_2 is stored as the four-carbon acid malate and used for photosynthesis during the day.

The pre-collected CO_2 is concentrated around the RuBisCO enzyme, increasing photosynthetic efficiency.

3. A is correct.

Autotrophs (i.e., *producers*) synthesize complex organic compounds (i.e., carbohydrates, fats, proteins) from simple substances in their environment, generally using energy from light (i.e., photosynthesis) or inorganic chemical reactions (i.e., chemosynthesis).

Autotrophs are the producers in a food chain (plants on land and algae in water), in contrast to heterotrophs that consume autotrophs. They can make their food and do not need living energy or organic carbon source.

Autotrophs can reduce carbon dioxide to make organic compounds.

Most autotrophs use water as the reducing agent, but some can use other hydrogen compounds.

4. B is correct.

Adenosine triphosphate (ATP) is a *nucleotide* composed of adenosine (A), ribose, and three phosphate groups.

ATP has three phosphate groups and is produced by enzymes (e.g., ATP synthase) from adenosine diphosphate (ADP) or adenosine monophosphate (AMP) and phosphate groups.

Mitochondria are the site of most ATP molecules produced during aerobic respiration.

5. C is correct.

Photosynthesis reaction:

$$6\ CO_2 + 6\ H_2O \rightarrow C_6H_{12}O_6 + 6\ O_2$$

The reactants are six carbon dioxide (CO_2) and six water (H_2O) molecules, converted by light energy captured by chlorophyll pigments into one sugar ($C_6H_{12}O_6$) and six oxygen (O_2) molecules.

Chloroplasts are the site of photosynthesis and are only in algae and plant cells.

Chloroplasts contain DNA and ribosomes and may have similarly evolved via endosymbiosis.

6. D is correct.

ATP is a nucleoside triphosphate with a nitrogenous base (adenine), ribose sugar, and triphosphate.

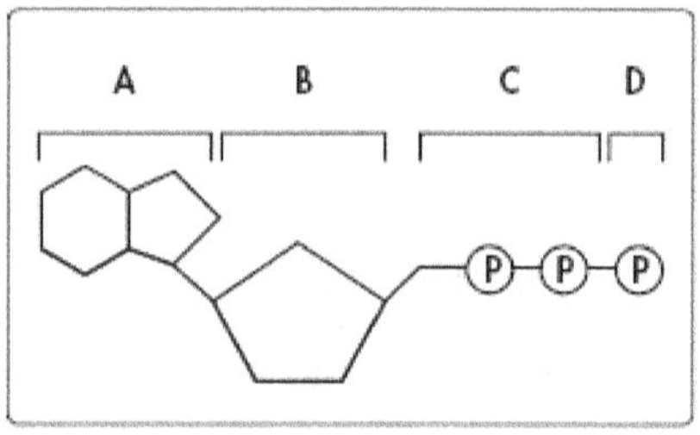

The complete structure represents ATP.

Structure A represents *adenine* (nitrogenous base).

Structure B represents *ribose.*

Structure C represents *diphosphate.*

Structure D represents *phosphate*

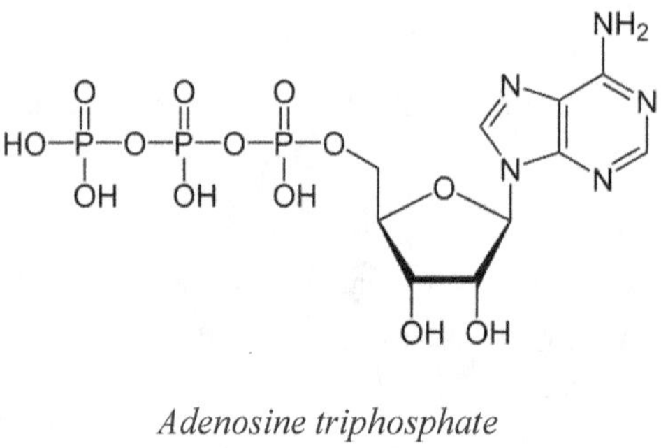

Adenosine triphosphate

7. A is correct.

Stroma is the colorless fluid surrounding the thylakoids within the chloroplast.

Within the stroma are *grana* (i.e., stacks of thylakoids) – the sub-organelles,
where photosynthesis begins before the chemical changes are completed in the stroma.

8. B is correct.

During photosynthesis, autotrophs produce carbohydrates (e.g., glucose).

Photosynthesis:

$$6\ CO_2 + 6\ H_2O \rightarrow C_6H_{12}O_6 + 6\ O_2$$

Reactants are six carbon dioxide (CO_2) and six water (H_2O) molecules, converted by light energy captured by chlorophyll pigments into one sugar ($C_6H_{12}O_6$) and six oxygen (O_2) molecules.

9. A is correct.

Stomata are composed of a pair of specialized epidermal cells known as *guard cells*.

Stomata regulate gas exchange (e.g., CO_2) and control water loss by changing the size of the stomatal pore.

10. E is correct.

Plants use photosynthesis to make food.

During photosynthesis, plants absorb photons (light energy) with their leaves. Plants use energy from the sun to convert water (H_2O) and carbon dioxide (CO_2) into glucose ($C_6H_{12}O_6$), six-carbon chain sugar.

Plants use glucose ($C_6H_{12}O_6$) for cellular energy to make biomolecules (e.g., cellulose and starch in plants; glycogen in animals).

Cellulose builds cell walls, while starch is stored in seeds and plant parts as a food source.

Chlorophylls *a* and *b* are photosynthetic pigments produced in chloroplasts of photosynthetic tissues of leaves.

Chlorophyll molecules are water-repelling because of the long phytol tail in the molecule.

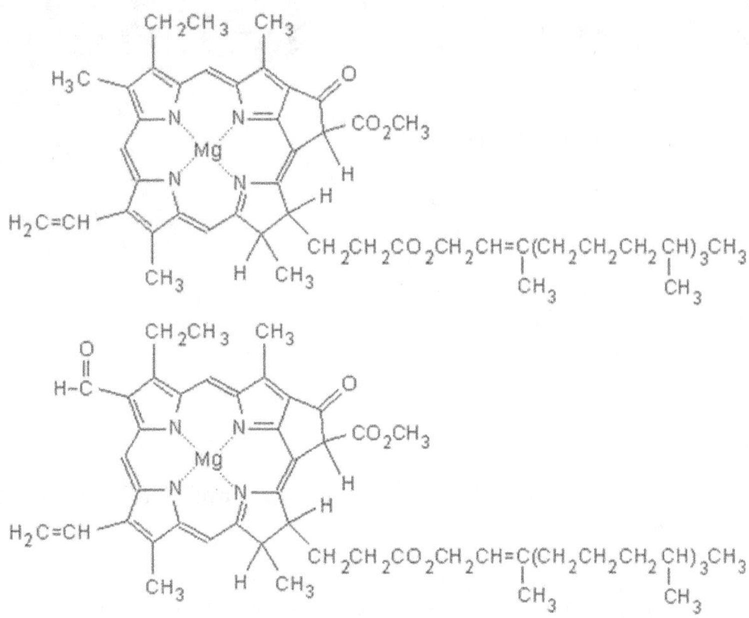

Photopigments chlorophyll a (top) and chlorophyll b (bottom) absorb energy during photosynthesis

continued...

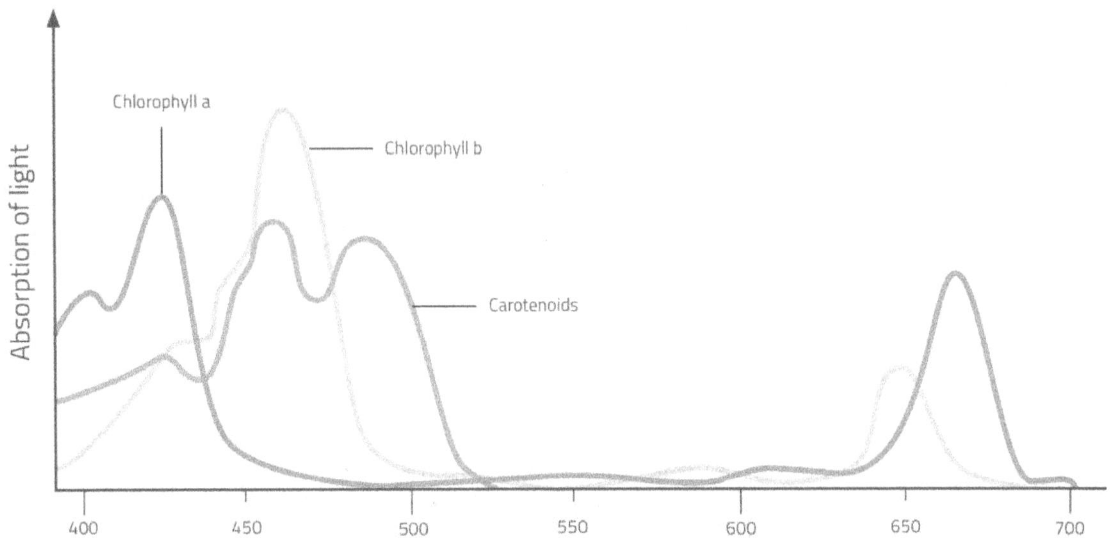

Photopigments absorb specific wavelengths of light with a visible light spectrum from 380 to 700 nm

11. C is correct.

Roots are used by plants to obtain most of their water.

12. C is correct.

Thylakoids are labeled in the figure.

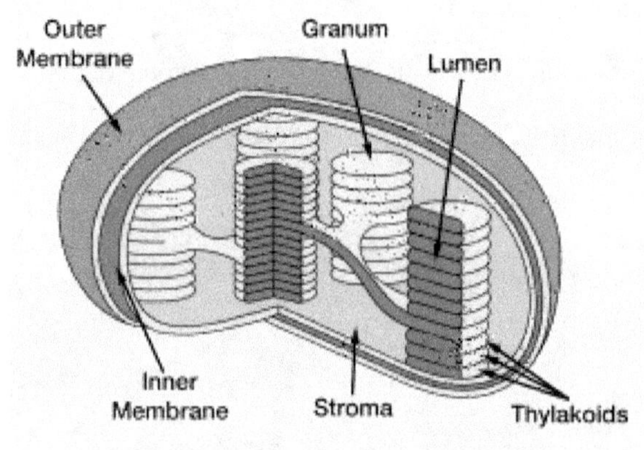

Chloroplast with internal structures

13. E is correct.

Thylakoids are membrane-bound compartments inside chloroplasts and cyanobacteria.

Thylakoids are the site of photosynthesis.

Thylakoids consist of a thylakoid membrane surrounding a thylakoid lumen.

Thylakoids stack as disks (i.e., grana).

14. B is correct.

Chlorophyll is within the thylakoid membrane.

15. A is correct.

Granum (pl. grana) is a stack of thylakoid discs. Chloroplasts can have from 10 to 100 grana.

Grana are connected by stroma thylakoids (or lamellae).

Grana thylakoids and stroma thylakoids are distinguished by their protein composition.

16. A is correct.

Electron carrier molecule in the figure is $NADP^+$.

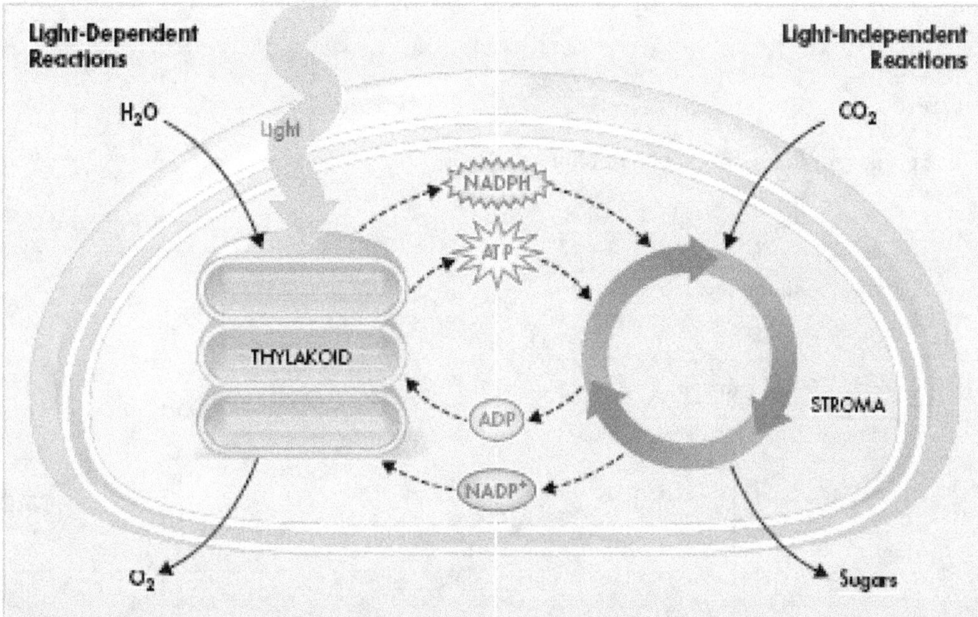

Chloroplast with light-dependent (ATP and NADPH) and light-independent reactions (sugars)

17. B is correct.

Low ATP production is likely if a shade-tolerant plant has minimal water intake.

18. E is correct.

High-energy electrons are highly reactive and are transported in the chloroplast by *electron carriers* ($NADP^+$).

19. D is correct.

Photosynthesis reaction:

$$6\ CO_2 + 6\ H_2O \rightarrow C_6H_{12}O_6 + 6\ O_2$$

Reactants are six carbon dioxide (CO_2) and six water (H_2O) molecules, converted by light energy captured by chlorophyll pigments into one sugar ($C_6H_{12}O_6$) and six oxygen (O_2) molecules.

20. A is correct.

Rate of photosynthesis *increases* and then levels off as the intensity of light increases.

21. A is correct.

In the redox reactions of photosynthesis, *electron transfer* is from $H_2O \rightarrow CO_2$

22. B is correct.

Steps in the light-dependent reaction include ATP synthase allowing H^+ ions to pass through the thylakoid membrane, high-energy electrons through the electron transport chain, and pigments in photosystem II absorbing photons of light.

23. E is correct.

Oxygen is the product of the light reaction of photosynthesis.

24. C is correct.

Photosystems I and II are in the *thylakoid membrane*.

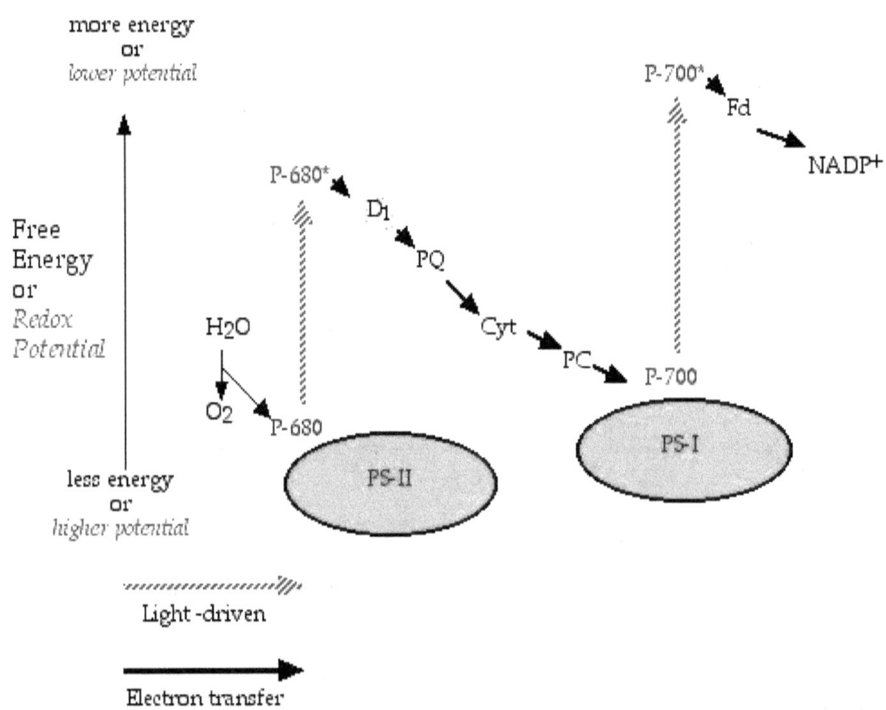

Photosystem II and I in the thylakoid membrane of the chloroplast. The light-driven reactions include electron promotions for P-680 and P-700. $NADP^+$ is the electron carrier.

25. B is correct.

Calvin cycle (i.e., C_3 cycle) is a series of biochemical redox reactions in the stroma of chloroplasts in photosynthetic organisms.

It is one of the light-independent reactions used for carbon fixation.

RuBisCO is the key enzyme of the cycle.

Photosynthesis occurs in two stages.

1) *Light-dependent reactions* capture the energy of light and use it to make energy storage and transport molecules ATP and NADPH.

2) *Light-independent reactions* (Calvin cycle) use energy from short-lived electronically excited electron carriers to convert carbon dioxide and water into organic compounds used by the organism (and animals that feed on it).

This set of reactions is *carbon fixation.*

Calvin cycle has four steps:

1) carbon fixation,

2) reduction phase,

3) carbohydrate formation, and

4) regeneration phase.

ATP and NADPH provide energy for the chemical reactions in this sugar-generating process.

Sunlight provides the energy for the reaction:

ATP + NADPH + carbon dioxide (CO_2) → sugar

diagram below...

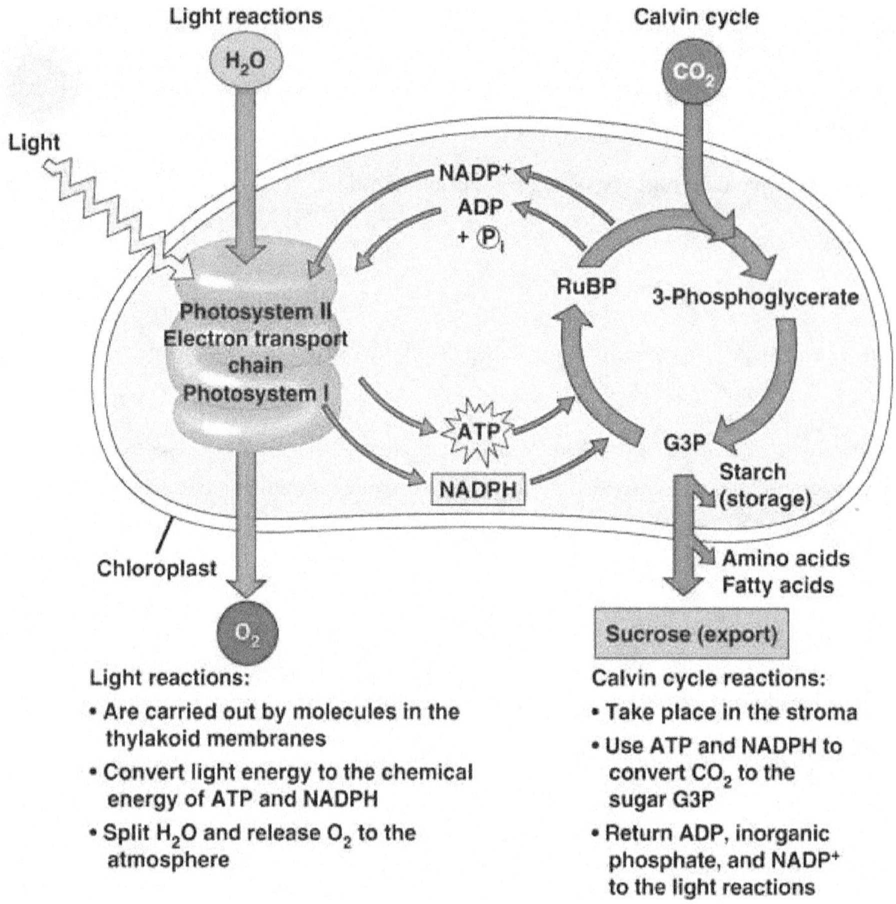

Products of light-dependent (ATP and NADPH) and light-independent reactions (sugars)

26. E is correct.

Electrons flow during photosynthesis in $H_2O \rightarrow NADP^+ \rightarrow$ Calvin cycle.

27. D is correct.

Light-dependent reactions of photosynthesis produce NADPH and ATP.

28. D is correct.

Calvin cycle occurs in the *stroma* of chloroplasts.

29. C is correct.

Blue-colored plants reflect light of blue wavelength (380 – 500 nm); visible light spectrum (380 – 700 nm).

30. A is correct.

Three components of an ATP molecule are adenosine (*base*), ribose (*sugar*), and three *phosphate* groups.

31. C is correct.

Wavelength *reflected* determines the color of an object.

32. A is correct.

Energy is *released* by the hydrolysis of ATP (triphosphate) into ADP (diphosphate).

33. E is correct.

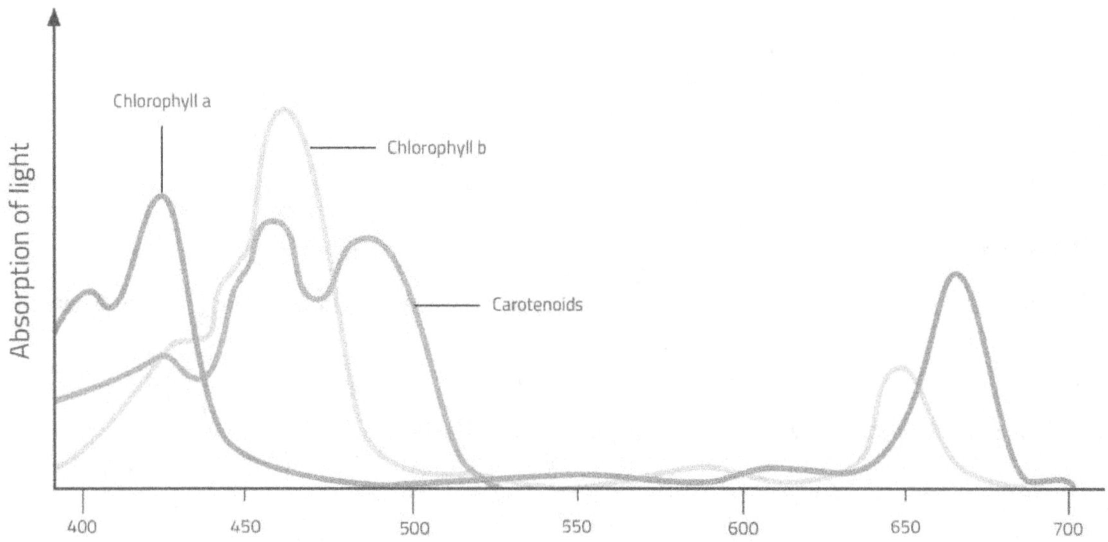

Photopigments absorb specific wavelengths of light with a visible light spectrum from 380 to 700 nm

34. C is correct.

Autotrophs (i.e., *producers*) synthesize complex organic compounds (i.e., carbohydrates, fats, proteins) from simple substances in their environment.

Autotrophs use energy from light (i.e., photosynthesis) or inorganic chemical reactions (i.e., chemosynthesis).

Autotrophs are the producers in a food chain (plants on land and algae in water), in contrast to heterotrophs that consume autotrophs.

Autotrophs can make their food and do not need living energy or organic carbon source.

Autotrophs can reduce carbon dioxide to make organic compounds.

Most autotrophs use water to reduce (i.e., a gain of electrons), but some can use other hydrogen compounds.

Phototrophs (i.e., green plants and algae), a type of autotroph, convert physical energy from sunlight into chemical energy in the form of reduced carbon.

35. A is correct.

Carotenoids reflect light with yellow, red, and orange wavelengths.

36. D is correct.

Mushrooms are *heterotrophs* and derive nutrients from *complex organic molecules*.

37. E is correct.

Plants use chlorophyll *a*, chlorophyll *b*, and carotenoid accessory pigments to absorb light over a broad spectrum of wavelengths.

38. B is correct.

The *sun* is the *ultimate energy source* for living organisms.

39. A is correct.

Photons are discrete packets of light.

40. B is correct.

Most plants appear green because chlorophylls do *not* absorb wavelengths of green light (500 – 560 nm).

41. A is correct.

Electromagnetic spectrum (ROY G BIV) has higher energy light with shorter wavelengths.

42. B is correct.

Thylakoids are membrane-bound compartments inside chloroplasts and cyanobacteria and the site of photosynthesis.

43. B is correct.

Energy from photons is transferred to electrons, raising them to a higher energy (i.e., excited) level.

44. C is correct.

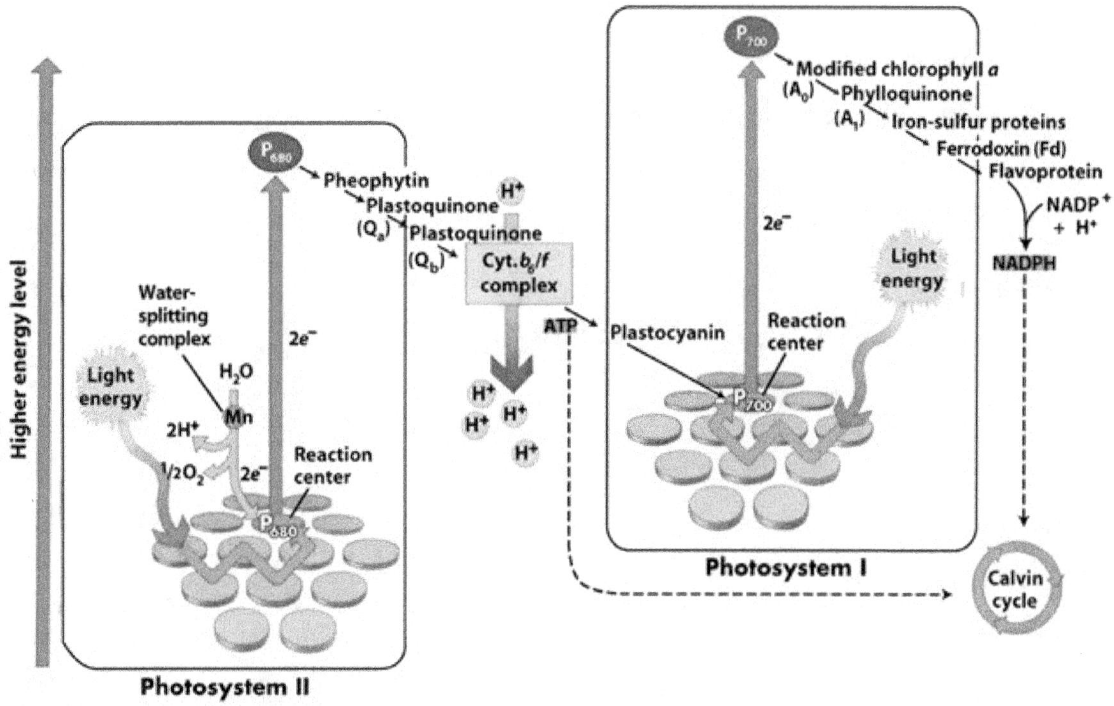

Photosystem II and I in the thylakoid membrane of the chloroplast. The light-driven reactions include electron promotions for P-680 and P-700 with $NADP^+$ as the electron carrier.

45. E is correct.

Molecules can *release energy* as fluorescence, heat, or light.

46. A is correct.

Photosynthesis reaction:

$$6\ CO_2 + 6\ H_2O \rightarrow C_6H_{12}O_6 + 6\ O_2$$

Reactants are six carbon dioxide (CO_2) and six water (H_2O) molecules, converted by light energy captured by chlorophyll pigments into one sugar ($C_6H_{12}O_6$) and six oxygen (O_2) molecules.

47. D is correct.

Photosystems are in the *thylakoid* membrane of chloroplasts.

48. A is correct.

Light-dependent reactions occur within *thylakoid membranes*.

49. E is correct.

Chlorophyll *a* is a form of chlorophyll used in photosynthesis, absorbing most energy from wavelengths of violet-blue and orange-red light.

Chlorophyll *a* reflects green and yellow light and contributes to the green color of many plants.

Photosynthetic pigment is essential for photosynthesis in eukaryotes and cyanobacteria because it is a primary electron donor in the *electron transport chain.*

50. C is correct.

Products of the *light-dependent reaction* are ATP, NADPH, and O_2 gas.

51. A is correct.

Calvin cycle has four steps:

carbon fixation,

reduction phase,

carbohydrate formation, and

regeneration.

ATP and NADPH provide energy for the chemical reactions in this *sugar-generating process*.

Sunlight provides energy for the reaction:

ATP + NADPH + carbon dioxide (CO_2) → sugar

52. B is correct.

Calvin cycle (i.e., C_3 cycle) is a series of biochemical redox reactions in the stroma of chloroplasts in photosynthetic organisms.

Calvin cycle is one of the light-independent reactions used for carbon fixation.

RuBisCO is the key enzyme of the Calvin cycle.

Photosynthesis occurs in two stages in a cell.

Light-dependent reaction captures energy from light and uses it to make the energy storage and transport molecules ATP and NADPH.

*Light-independen*t Calvin cycle uses energy from short-lived electronically excited electron carriers to convert carbon dioxide and water into organic compounds used by the organism (and animals that feed on it).

This set of reactions is *carbon fixation*.

53. D is correct.

Water is the source of *oxygen* and electrons released by a photosystem.

54. C is correct.

Light-independent reaction of photosynthesis used for carbon fixation is the *Calvin cycle.*

55. B is correct.

Water is the source of oxygen and *electrons* released by a photosystem.

56. E is correct.

Electron transport chain occurs in light reactions of photosynthesis (in chloroplasts) and *cellular respiration* (in mitochondria).

57. C is correct.

H^+ gradient is formed across the *thylakoid membranes* of chloroplasts during photosynthesis.

Notes for active learning

Notes for active learning

Notes for active learning

Section II: Molecular Biology of Cells

Molecular Biology of Eukaryotes – Detailed Explanations

1. A is correct.

5'-CCCC-3' and 5'-AAAA-3'primers should be used to amplify the DNA shown via PCR.

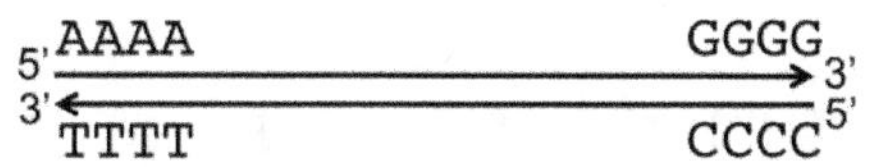

Polymerase chain reaction (PCR) requires the sequence at the ends of the fragment to be amplified.

From the ends, primers use complementary base pairing to anneal the target fragment and permit amplification.

Regions between ends are not required as the parent strands are the template used by the DNA polymerase.

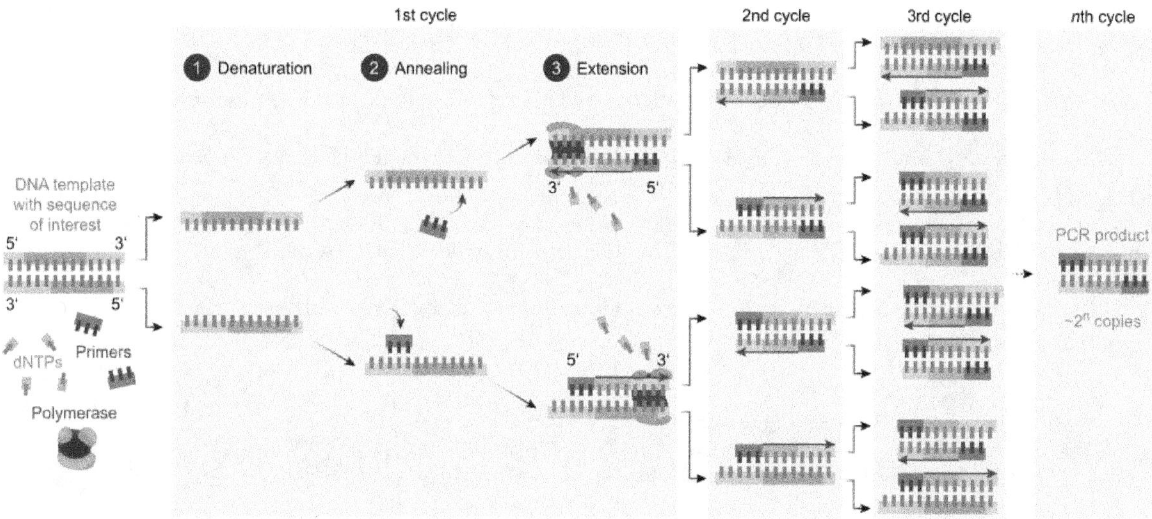

Polymerase chain reaction (PCR) with exponential product amplification during each cycle. Step one heats DNA to separate the two strands of the double helix. Step 2 cools the sample for annealing by complementary primers. Step 3 is chain elongation for the synthesis of DNA by polymerase anchored to the primers.

2. E is correct.

Proteins are the major phenotypic expression of a genotype.

3. C is correct.

Dideoxy (without ~OH at 2' and 3' positions) cytosine is used in DNA sequencing to cause termination when the template strand is G.

4. A is correct.

Adding a 3' poly-A tail to mRNA (not proteins) is a *post-transcription* event.

Post-transcription events include adding a 5' cap and splicing exons (removing introns) from RNA molecules.

5. C is correct.

Centromeres are chromosomal regions of highly coiled DNA (i.e., *heterochromatin*) at the point of attachment of the sister chromatids, replicated during the S phase of interphase.

6. D is correct.

Genes provide hereditary information for the organism, but the environment (to a degree) influences the phenotypic pattern of proteins resulting from gene expression.

Genetic and environmental factors interact to produce the phenotype. These are examples of the environment affecting gene activation.

Shivering is an example of a change in the environment causing a physiological (i.e., behavioral) change to maintain homeostasis. Homeostasis is achieved because shivering generates heat when muscles contract rapidly.

7. C is correct.

Library refers to DNA molecules inserted into cloning vectors (e.g., bacterial plasmid).

Complementary DNA (cDNA) is created from mRNA in a eukaryotic cell using reverse transcriptase.

In eukaryotes, a poly-A tail (i.e., a sequence of 100-250 adenosine (A) nucleotides) distinguishes mRNA from tRNA and rRNA.

Only mRNA has the poly-A tail bind (i.e., mRNA is isolated from the other RNA molecules).

Poly-A tail can be used as a primer site for *reverse transcription* (i.e., mRNA → cDNA).

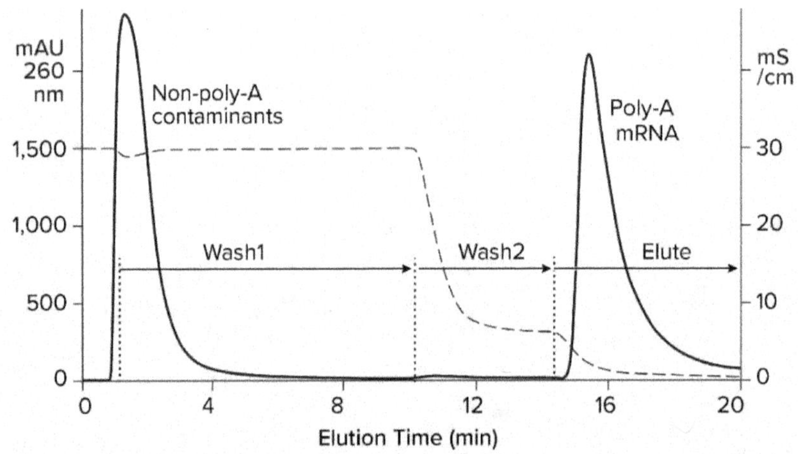

mRNA is purified by column chromatography using oligomeric dT nucleotide resins

mRNA is isolated by oligo-dT and used as a complementary primer to bind the poly-A tail and extend the primer molecule by reverse transcriptase to create the cDNA strand.

The original mRNA templates are removed using RNAse, leaving a *single-stranded* cDNA (sscDNA).

cDNA is converted into double-stranded DNA by DNA polymerase for double-stranded cDNA.

cDNA is cloned in bacterial plasmids.

8. A is correct.

Quaternary protein structure requires two or more polypeptide chains whereby different genes specify each polypeptide.

Complex polysaccharides are linked to monomers of sugar. Complex polysaccharides include sugar monomers such as lactose (e.g., glucose and galactose).

The presence of more than one monomer requires additional genes.

9. E is correct.

Genomic library is a set of clones representing a given organism's entire genome. Among the tools making recombinant DNA technology possible are *restriction enzymes.*

Restriction endonucleases (i.e., *restriction enzymes*) cut DNA at specific nucleotide sequences, often inverted repeat sequences (i.e., palindromes that read the same if the DNA strand is rotated 180°).

After cutting, the ends of some restriction-digested double-stranded DNA fragments have several nucleotide overhangs of single-stranded DNA as "sticky ends."

Restriction enzymes cut similarly, so the sticky end from one fragment anneals via hydrogen bonding with the sticky end from another fragment cut by the same enzyme.

After annealing, DNA ligase covalently closes the plasmid into circular DNA for transformation into bacteria.

10. D is correct.

Attachment of glycoprotein side chains is a post-translational modification.

Rough endoplasmic reticulum (RER) and the *Golgi apparatus* are two organelles modifying proteins as a post-translational event.

Lysosomes are organelles with low pH that function to digest intracellular molecules.

11. A is correct.

Telomerase (i.e., enzyme ending in *~ase*) restores the ends of the DNA in a chromosome.

Telomeres are maintained by *telomerase*, an enzyme that (during embryogenesis) adds repeats to chromosomal ends. A measure of telomerase activity in adults indicates one marker for cancer activity (i.e., uncontrolled cell growth) within the cell.

DNA polymerase cannot fully replicate the 3' DNA end, resulting in shorter DNA with each division.

The new strand synthesis mechanism prevents the loss of the DNA coding region.

New strand synthesis prevents the loss of the terminal DNA at the coding region during replication and represents telomere function.

Telomeres are at chromosomal ends and consist of nucleotide repeats that provide stability and prevent the loss of the ends of the DNA coding region during DNA replication.

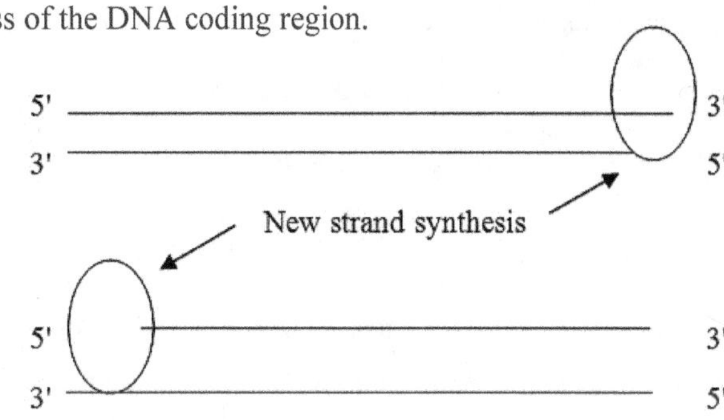

12. B is correct.

Unlike a genomic library that reflects the entire DNA of the organism, a *cDNA library* uses mRNA from *expressed genes after processing* (i.e., removal of introns and ligating of exons).

13. E is correct.

Central dogma of molecular biology proposes that information flow in the cell is unidirectional.

Schematically,

DNA → RNA → protein

14. C is correct.

Alternative splicing occurs in the nucleus with three stages of RNA processing:

1) adding a G-modified cap to the 5' end of the primary transcript,

2) adding a 3' poly-A tail, and

3) removing introns (intervening sequences) and ligating exons (expressed sequences).

This process cleavages hnRNA to make corresponding mRNAs transported in the cytoplasm.

15. D is correct.

Ribosomal RNA (rRNA) binds the small and large ribosomal subunits (i.e., 30S & 50S in prokaryotes and 40S & 60S in eukaryotes) to create the functional ribosome (i.e., 70S in prokaryotes and 80S in eukaryotes).

Ribosomes are complex assemblies of RNA molecules and several associated proteins. rRNAs are the only RNA molecules (others being mRNA and tRNA) synthesized in the nucleolus (within the nucleus).

16. A is correct.

E. coli RNA polymerase synthesizes RNA in the 5' to 3' direction and copies a DNA template.

17. B is correct.

Introns (intervening sequences) vary in size and number among different genes.

18. E is correct.

I: eukaryotes *splice* hnRNA (*primary transcript*) by *removing introns* (intervening sequences) and *ligating exons* (expressed sequences) in the nucleus before exporting the modified mRNA transcript to the cytoplasm for translation.

A poly-A tail to the mRNA is added before export from the nucleus.

II: after protein synthesis in the ER, they may be modified in the Golgi apparatus (e.g., glycosylation), but this is different from the required *splicing* of mRNA molecules (i.e., removal of introns) before the mRNA can pass through the nuclear pores and enter the cytoplasm.

III: *prokaryotic ribosomes* (30S small & 50S large subunit = 70S complete ribosome) are smaller than *eukaryotic ribosomes* (40S small & 60S large subunit = 80S complete ribosome).

19. C is correct.

Deacetylation of histones associated with the retrotransposon and methylation of retrotransposon DNA decreases the transcription of retrotransposons.

20. B is correct.

Central dogma of molecular biology refers to the direction of genetic information flow.

DNA → RNA → protein

Central dogma of molecular biology is violated by *retroviruses* with RNA genomes and reverse transcriptase to copy viral RNA into DNA within the infected host cell.

21. A is correct.

miRNA (*interference mRNA*) is generated from the cleavage of double-stranded RNA. miRNA is a small non-coding RNA molecule (about 22 nucleotides) in plants, animals, and some viruses, which functions in transcriptional and post-transcriptional regulation of gene expression.

miRNA hybridizes to mRNA and inactivates its ability as a template for translation into protein.

22. C is correct.

DNA polymerase cannot fully replicate the 3' DNA end, resulting in shorter DNA with each division. The new strand synthesis mechanism prevents the loss of the DNA coding region.

The process of new strand synthesis prevents the loss of the terminal DNA at the coding region during replication by *telomeres*. Telomeres are at chromosomal ends and consist of nucleotide repeats that provide stability and prevent the loss of the ends of the DNA coding region during DNA replication.

Telomeres are generated by *telomerase*, an enzyme that (during embryogenesis) adds repeats to chromosomal ends. A measure of telomerase activity in adults indicates one marker for cancer activity (i.e., uncontrolled cell growth) within the cell.

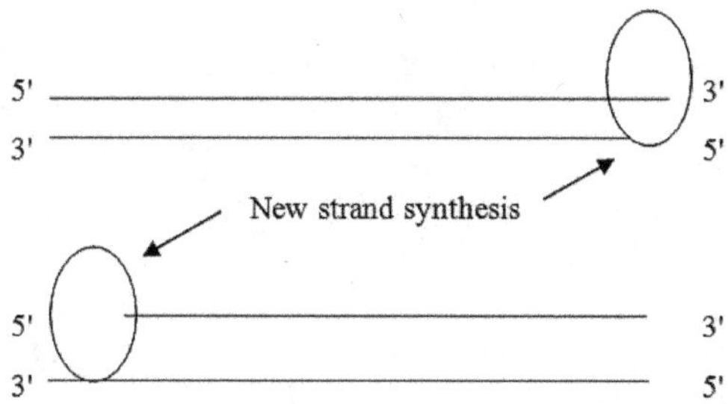

A: *kinetochore* is a protein collar around the centromere on the chromosome as the site of attachment of the spindle fiber to the chromosome during cell division.

Kinetochore attaches to the spindle fibers (microtubulin) and is anchored to chromosomes by centromeres. The other end of the spindle fibers is attached at the poles (centromere region) of the cell to the centrioles.

B: *centrosome* (*~some* means body) is a region at the poles of the cell giving rise to the asters (microtubules projecting past the centrioles) of the centrioles.

D: *centromeres* are on the chromosomes consisting of highly coiled DNA (i.e., heterochromatin) at the point of attachment of the sister chromatids.

23. B is correct.

Core particle of 8 *histones* (i.e., positively charged DNA-binding proteins) is the *nucleosome* (i.e., octamer).

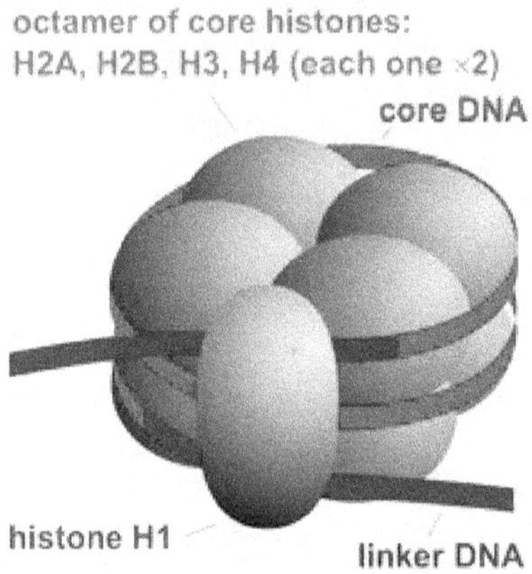

Nucleosome of 8 histones joined by a linker region with histone H1 permit efficient coiling of DNA

24. C is correct.

miRNA (or *interference RNA*) is a small non-coding RNA molecule (~22 nucleotides) in plants, animals, and some viruses, which functions in transcriptional and post-transcriptional regulation of gene expression.

miRNA hybridizes to mRNA and inactivates its ability to be used as a template for protein translation.

25. A is correct.

Promoter is the region of DNA in prokaryotes to which *RNA polymerase binds* most tightly.

Promoter is a region of 100–1000 base pairs long on the DNA that initiates gene transcription.

Promoters are nucleotides near the genes they transcribe, on the same strand, and upstream on the DNA (towards the 3' region of the antisense strand – template and non-coding strands).

Promoters contain specific DNA sequences and response elements that provide a secure initial binding site for RNA polymerase and transcription factors proteins.

Transcription factors have specific activator (or repressor) sequences of nucleotides that attach to specific promoters and regulate gene expressions.

In bacteria, the promoter is recognized by RNA polymerase and an associated sigma factor (i.e., a protein needed only for initiation of RNA synthesis), which are often brought to the promoter DNA by an activator protein's binding to its own nearby DNA binding site.

In eukaryotes, the process is complicated, and at least seven factors are necessary for binding an RNA polymerase II to the promoter.

26. E is correct.

Genomic libraries reflect the entire DNA of the organism and, unlike a cDNA library, do not use mRNA, which corresponds to expressed genes after processing (i.e., removal of introns).

Retrotransposons are genetic elements that can amplify themselves in a genome and are ubiquitous components of the DNA of many eukaryotic organisms.

27. D is correct.

tRNA has a *cloverleaf secondary structure* and is the smallest single-stranded RNA.

tRNA contains the *anticodon* and delivers individual amino acids to the growing peptide chain based on the *codon* specified by mRNA.

A: hnRNA (*heteronuclear RNA* or *primary transcript*) has not yet undergone post-transcriptional modification (i.e., 5' Guanine cap, removal of introns and ligation of exons, and the addition of a 3' poly-A tail).

Once three modifications occur, the hnRNA becomes a mature mRNA and is transported through the nuclear pores (in the nuclear envelope).

mRNA enters the cytoplasm for translation into proteins.

B: mRNA is a single-stranded RNA molecule from its genesis and is shorter than hnRNA (i.e., the primary transcript before post-transcriptional processing).

mRNA does not contain introns (exons are joined) and has a 5' G-cap and a 3' poly-A tail.

C: rRNA is synthesized in the *nucleolus* (organelle in nucleus) and is the most abundant single-stranded RNA molecule.

rRNA (and associated proteins) are necessary for *ribosome* assembly in protein synthesis (i.e., translation).

E: miRNA (*interference mRNA*) is generated from the cleavage of double-stranded RNA.

miRNA is a small non-coding RNA molecule (about 22 nucleotides) in plants, animals, and some viruses, which functions in transcriptional and post-transcriptional regulation of gene expression.

miRNA hybridizes to mRNA and inactivates its ability as a template for translation into protein.

28. B is correct.

Inversion is a chromosomal rearrangement when a chromosome segment is reversed and occurs when a single chromosome undergoes breakage and rearrangement.

Inversions do not cause abnormalities if rearrangement is balanced with no extra or missing DNA.

Heterozygous individuals for an inversion have increased production of abnormal chromatids resulting from crossing over within inversion spans.

Heterozygous inversions lead to *lowered fertility* due to the production of *unbalanced gametes*.

29. E is correct.

Nucleolus is the organelle within the nucleus responsible for synthesizing ribosomal RNA (rRNA).

A: *Golgi apparatus* is the organelle responsible for processing, packaging, and distributing proteins.

B: *lysosomes* are organelles with low pH that digest intracellular molecules.

C: *mitochondrion* is a double-membraned organelle where the Krebs (TCA) cycle, electron transport, and oxidative phosphorylation occur.

30. D is correct.

AG mutant flowers not having reproductive organs supports that the AG gene is vital for forming reproductive flower organs in *Arabidopsis*.

31. C is correct.

Glycocalyx refers to extracellular material, such as glycoprotein produced by bacteria, endothelial, epithelial, and other cells.

Glycocalyx consists of several carbohydrate moieties of membrane glycolipids and glycoproteins.

Glycocalyx is composed of a *negatively charged* network of proteoglycans, glycoproteins, and glycolipids on the apical surface of endothelial cells.

Carbohydrate portions of glycolipids on plasma membrane surfaces contribute to cell-cell recognition, communication, and intracellular adhesion.

Glycocalyx plays a significant role in endothelial vascular tissue, including the modulation of red blood cell volume in capillaries and a principal role in the vasculature to maintain plasma and vessel wall homeostasis.

Glycocalyx is on the apical surface of vascular endothelial cells that line the lumen and includes a wide range of enzymes and proteins that regulate leukocyte and thrombocyte adherence.

32. A is correct.

Chromatin (comprising chromosomes) is the combination of DNA and proteins that comprise the genetic contents within the nucleus of a cell.

Primary functions of *chromatin*:

1) package DNA into a smaller volume

2) strengthen DNA for mitosis

3) prevent DNA damage

4) control gene expression (i.e., RNA polymerase binding) and DNA replication

continued...

Three levels of *chromatin organization*:

1) DNA wraps around histone proteins, forming nucleosomes (i.e., *beads on a string*) in uncoiled euchromatin DNA;

2) multiple histones wrap into a 30 nm fiber consisting of nucleosome arrays in their most compact form (i.e., heterochromatin);

3) higher-level DNA packaging of the 30 nm fiber into the metaphase chromosome (during mitosis and meiosis).

Tandem repeats occur in DNA when a pattern of nucleotides is repeated, and the repetitions are adjacent.

For example, ATCCG ATCCG ATCCG, whereby the sequence ATCCG is repeated three times.

Tandem repeats may arise from errors during DNA replication whereby the polymerase retraces its path (over a short distance) and repeats the synthesis of the same template region of DNA.

33. E is correct.

DNA topoisomerases are nuclear enzymes that regulate DNA supercoiling.

Topoisomerases are essential in DNA replication, transcription, chromosome segregation, and recombination.

Cells have two topoisomerases:

type I, which makes single-stranded cuts in DNA, and

type II enzymes, which cut and pass double-stranded DNA.

Gyrase (a subset of topoisomerase II) creates double-stranded breaks between the DNA backbone and relaxes DNA supercoils by unwinding the nicked strand around the other.

B: *helicase* unwinds DNA and induces severe supercoiling in the double-stranded DNA when hydrolyzing the hydrogen bonds between the complementary base pairs and exposing the bases for replication.

C: *DNA polymerase I* is in prokaryotes and participates in excision repair with 3'-5' and 5'-3' exonuclease activity and processing of Okazaki fragments generated during lagging strand synthesis

D: *DNA polymerase III* is in prokaryotes and is the primary enzyme in DNA replication.

34. B is correct.

3' OH of cytosine attacks the phosphate group on the guanine.

35. A is correct.

Introns are intervening sequences in the primary transcript but are excised when RNA is processed into mRNA.

Processed mRNA consists of *exons ligated* (i.e., joined).

Lariat structures are the protein scaffolding used for the splicing of the hnRNA (i.e., primary transcript) into mRNA as the introns are removed, and the exons are ligated.

36. B is correct.

Histone H1 is not in the eight-particle core of a *nucleosome*.

Histone H1 is in the linker regions of about 50 nucleotides between the nucleosomes.

Nucleosome is an *octamer* of the core histones: two H2A, two H2B, two H3, and two H4 histones.

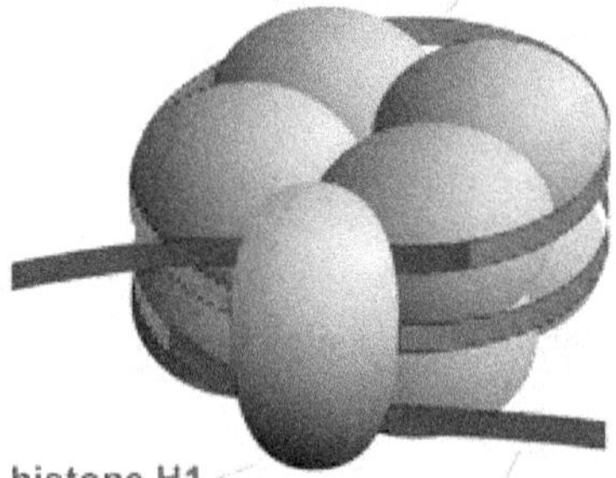

Nucleosome of 8 histones joined by a linker region with histone H1 permit efficient coiling of DNA

37. E is correct.

Cellulose comprises the plant cell wall and is not a chemical component of a bacterial cell wall (*peptidoglycan*).

38. D is correct.

Combinatorial control of gene transcription in eukaryotes is when the presence (or absence) of combinations of transcription factors is required.

39. D is correct.

Once synthesized via transcription, RNA molecules undergo three steps for post-transcriptional processing:

1) 5' G cap increases RNA stability and resistance to degradation in the cytoplasm;

2) 3' poly-A-tail functions as a *molecular clock*, and

3) introns (i.e., non-coding regions) are removed, and exons (i.e., coding regions) are spliced (joined).

40. B is correct.

The second round of transcription can begin before the initial transcript is completed.

41. A is correct.

SP1 is a *transcription factor* that binds nucleic acids.

Therefore, SP1 binds RNA and DNA. Select an organelle that lacks RNA or DNA.

Golgi apparatus is the organelle responsible for processing, packaging, and distributing proteins.

Golgi is not composed of nucleic acids, nor does it process them.

Since no nucleic acids exist within the Golgi apparatus, the transcription factor SP1 would not bind to it.

B: *mitochondrion* is the organelle bound by a double membrane, where the Krebs cycle, electron transport, and oxidative phosphorylation occur.

Mitochondria are the site of most ATP molecules produced during aerobic respiration.

Mitochondria (via the endosymbiotic theory) contain DNA (similar in size and composition to bacterial DNA), and the SP1 transcription factor would bind.

C: *nucleolus* is an organelle within the nucleus where rRNA synthesis occurs.

Thus, SP1 would bind to a nucleic acid (rRNA) within the nucleolus.

D: *ribosomes* contain rRNA and assembled proteins as the translation site (mRNA is converted into proteins).

42. E is correct.

Spliceosome is a large and complex molecular apparatus assembled from snRNP and protein complexes.

Spliceosomes remove introns from pre-mRNA primary transcript (i.e., heteronuclear RNA or hnRNA); catalyze the removal of introns and ligation of exons.

snRNA is a component of the snRNP and provides specificity by *recognizing* the sequences of critical splicing signals at the 5' and 3' ends and branch-site of introns.

Each spliceosome comprises five small nuclear RNAs (snRNA) and a range of associated protein factors.

The spliceosome occurs anew on each hnRNA (pre-mRNA).

hnRNA contains specific sequence elements recognized during spliceosome assembly; the 5' end splice, the branch point sequence, the polypyrimidine (i.e., cytosine and uracil) tract, and the 3' end splice site.

43. B is correct.

Polynucleotides (e.g., DNA) are the only of the four macromolecules (i.e., nucleic acids, proteins, lipids, carbohydrates) repaired rather than degraded.

DNA repair is essential for maintaining cell function and is performed by biological repair systems (i.e., p53 tumor repressor protein).

Uncontrolled cell growth via disruption to the cell cycle regulation may occur by somatic cell mutations.

44. D is correct.

Poly-A tail of RNA is enzymatically added soon after transcription (DNA → hnRNA) is completed.

45. E is correct.

H1 histones are in about 50 nucleotides between the nucleosomes in the linker regions.

Nucleosome is an *octamer* of the core histones: two H2A, two H2B, two H3, and two H4 histones.

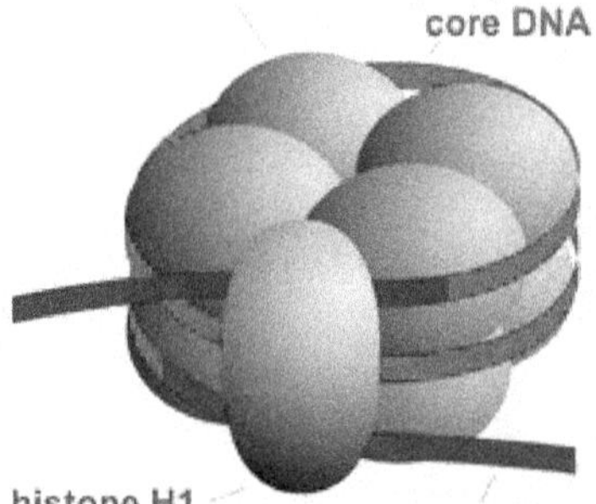

Nucleosome of 8 histones joined by a linker region with histone H1 permit efficient coiling of DNA

46. B is correct.

Splicing process involves snRNA which consists of sugar ribose (like other RNA molecules).

RNA is labile (unstable) and can function as a nucleophile because of the 2'–OH.

47. C is correct.

Chromatin is composed of DNA and histone proteins.

DNA has an overall negative charge because of the negative oxygen (i.e., single-bonded with seven valence electrons) attached to the phosphate group attached to the deoxyribose sugar.

For electrostatic interactions, histones must be *positively charged.*

Histones have a high concentration of basic (positive charge) amino acids (e.g., lysine and arginine).

48. C is correct.

Nucleic acids (e.g., DNA, RNA) elongate via *nucleophilic attack* by 3'–OH of nascent (i.e., growing) strand.

3'–OH undergoes nucleophilic attack on the phosphate group closest to the sugar in the incoming nucleotide.

2'–OH is in RNA (i.e., ribose) compared to a 2'–H for DNA (i.e., deoxyribose, where *deoxy* signifies the absence of O as in OH).

49. A is correct.

Heterochromatin is tightly packed DNA.

Heterochromatin consists of genetically inactive satellite sequences (i.e., tandem repeats of noncoding DNA).

Euchromatin is loosely coiled regions of DNA actively engaged in gene expression because RNA polymerase binds the relaxed region of DNA.

Centromeres and *telomeres* are *heterochromatic.*

Barr body of the second inactivated X-chromosome in females is *heterochromatic.*

50. D is correct.

RNA of the prokaryote does *not* undergo post-transcriptional modifications:

addition of 5' cap,

removal of introns and ligation of exons, and

addition of a 3' poly-A tail), as does hnRNA (*primary transcript*) in eukaryotes.

Prokaryotes lack a nucleus, so RNA synthesis (i.e., *transcription*) coincides with protein synthesis (i.e., *translation*) in the cytosol.

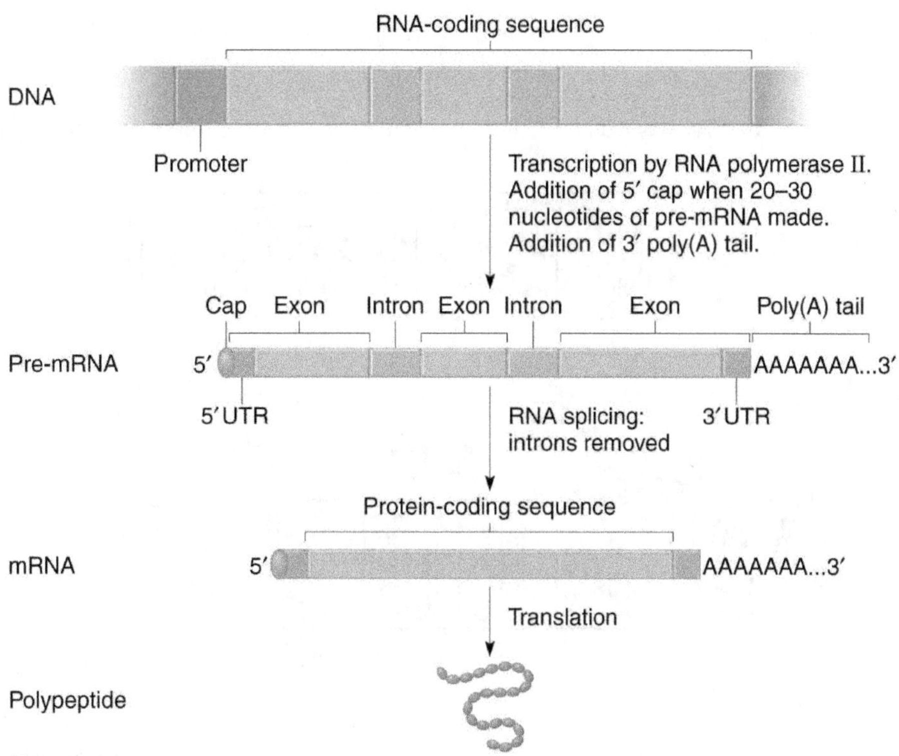

Gene expression with transcription, RNA processing, and mRNA translation into protein. Nucleotide sequences on chromosomes are transcribed and processed into mRNA within the nucleus by joining exons (removing introns), adding a 5' cap and a poly-A tail.

51. A is correct.

Recombinant DNA is nucleotides formed in the laboratory through genetic recombination (e.g., molecular cloning) to combine genetic material from multiple sources, creating sequences not in biological organisms.

Recombinant DNA is possible because DNA molecules from organisms share the same chemical structure. They differ in the nucleotide sequence within that identical overall structure.

52. E is correct.

Transformation (i.e., introducing foreign DNA into cells) is successful when foreign DNA is integrated into another organism's genome.

Once integrated, it replicates and divides with the host cell.

53. B is correct.

Restriction enzymes are *endonucleases* that cut DNA at specific *internal sequences*, usually inverted repeat sequences (i.e., palindromes that read the same if the DNA strand is rotated 180°).

54. B is correct.

Restriction enzymes cut DNA at specific sequences.

Endonucleases (i.e., enzyme cutting within the strand) *cleave double-stranded DNA* (dsDNA) between the nucleotides adenosine (A) and guanine (G) palindrome sequences.

Sticky ends are overhangs of TTAA as reading 5' to 3' (by convention, reading top strand left to right).

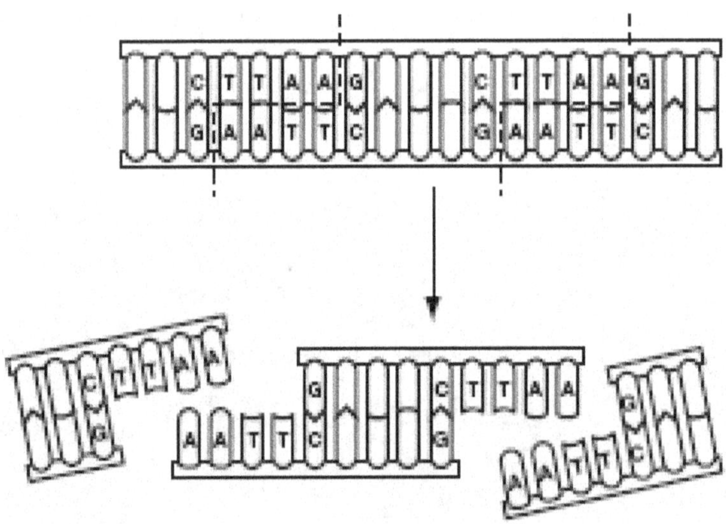

Notes for active learning

Notes for active learning

Biomolecules – Detailed Explanations

1. A is correct.

The primary structure of a protein is the linear sequence of amino acids.

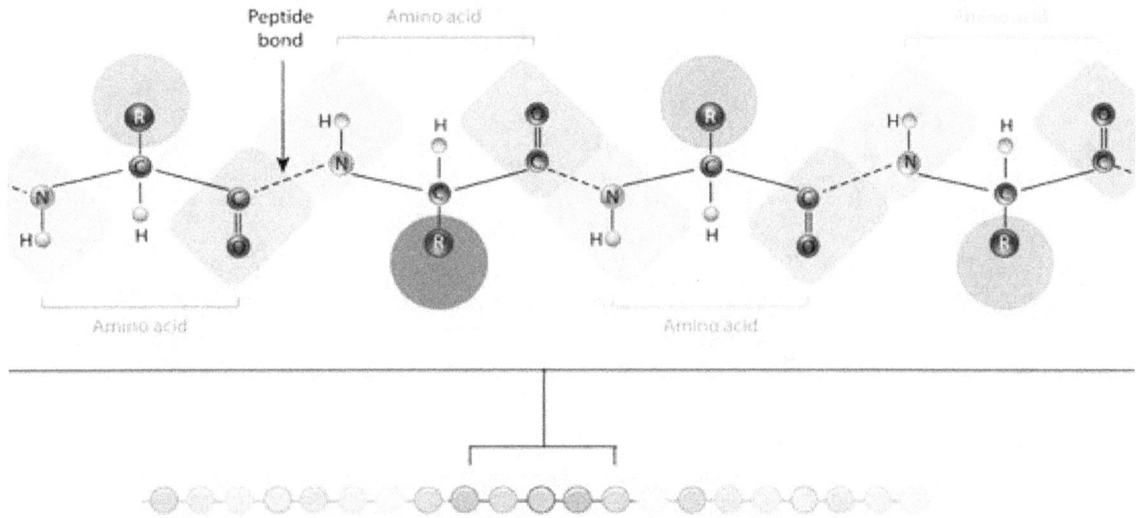

2. B is correct.

All the molecules in the body are divided into four categories. These biomolecular categories are carbohydrates, lipids, nucleic acids, and proteins.

3. A is correct.

Amino acids are linked in peptides by amide bonds. The peptide forms in a condensation reaction *via* dehydration (loss of water) when the lone pair on the nitrogen of an amino group of one amino acid makes a nucleophilic attack on a carbonyl carbon.

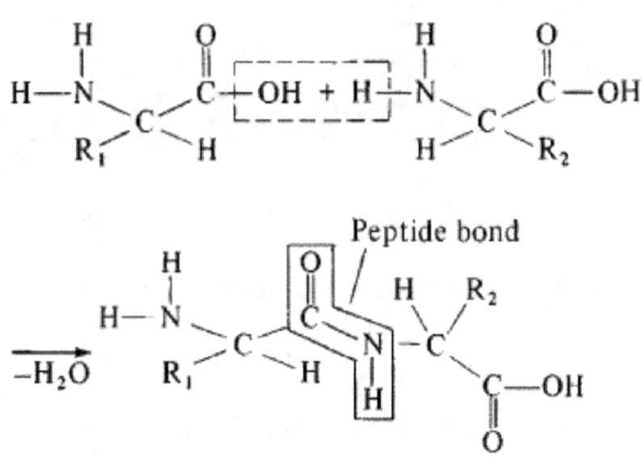

Peptide bonds form from the condensation reaction of 2 amino acids

4. B is correct.

The *alpha helix structure* is one of the two common types of secondary protein structure. The alpha helix is held together by hydrogen bonds between every N–H (amino group) and the oxygen of the C=O (carbonyl) in the next turn of the helix; four amino acids along the chain.

The typical alpha helix is about 11 amino acids long.

The other type of secondary structure is the *beta-pleated sheet.*

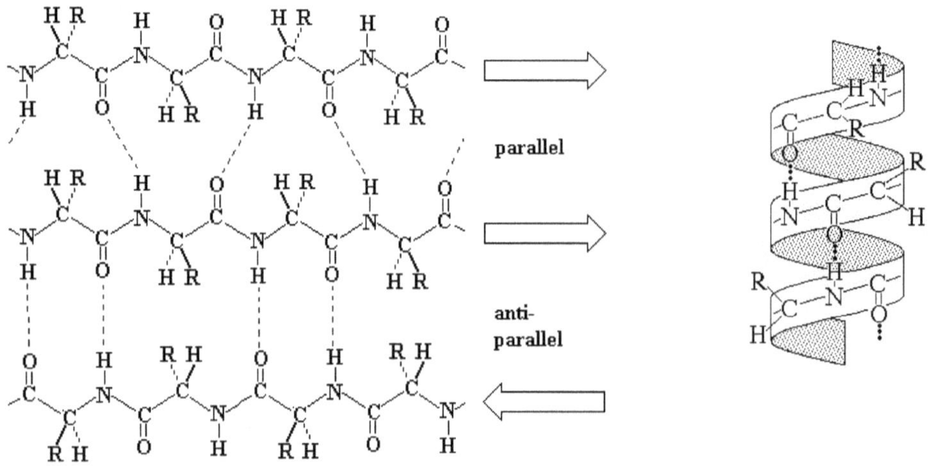

Beta pleated sheets are parallel or anti-parallel (a reference to the amino terminus).

5. C is correct.

The primary structure of a protein is the amino acid sequence, which is formed by covalent peptide linkages.

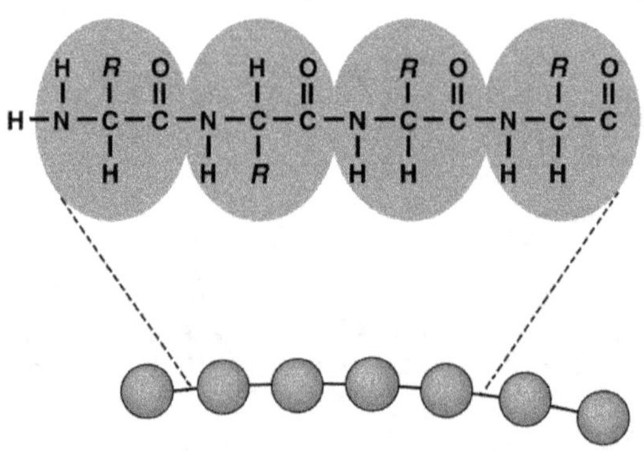

The amino acids (circles) are joined by covalent peptide bonds (lines)

A: only proteins containing more than one peptide subunit have a quaternary structure.

B: proteins are *denatured* by heating, and they lose their conformation above 35-40 °C, not retained.

D: many proteins contain more than one peptide chain (i.e., have a quaternary structure).

6. B is correct.

Amino acids are the building blocks of proteins.

Humans need 20 amino acids, some are made by the body (i.e., nonessential), and others must be obtained from the diet (i.e., essential).

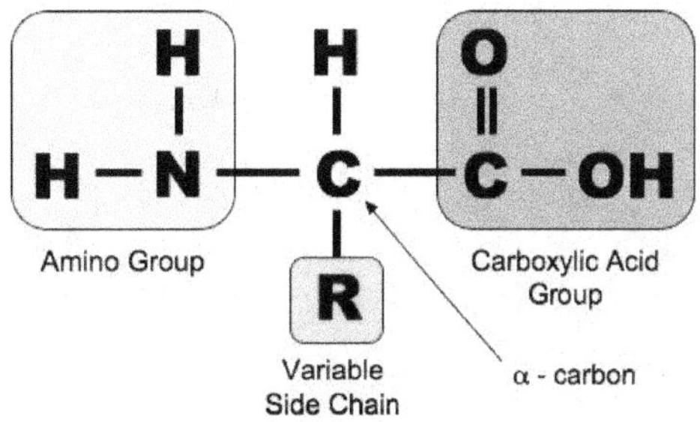

Amino acids contain an amine group, a carboxylic acid, an α-carbon, and an R group.

The following table is shown not for memorization but for identifying characteristics (e.g., polar, nonpolar) for the side chains.

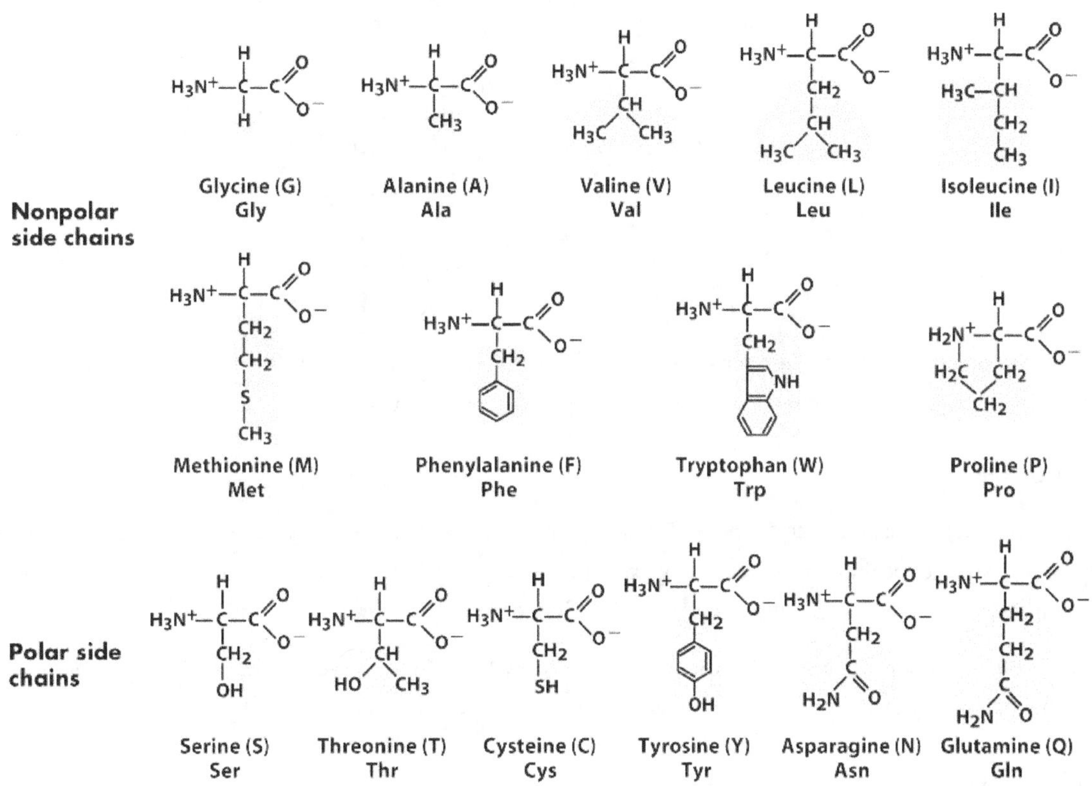

continued...

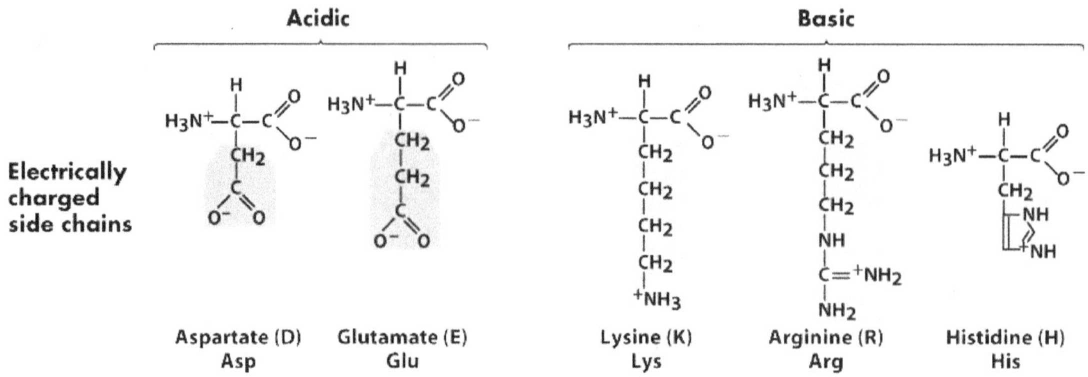

The 20 naturally occurring amino acids.

7. C is correct.

Amino acids are the basic building blocks for proteins.

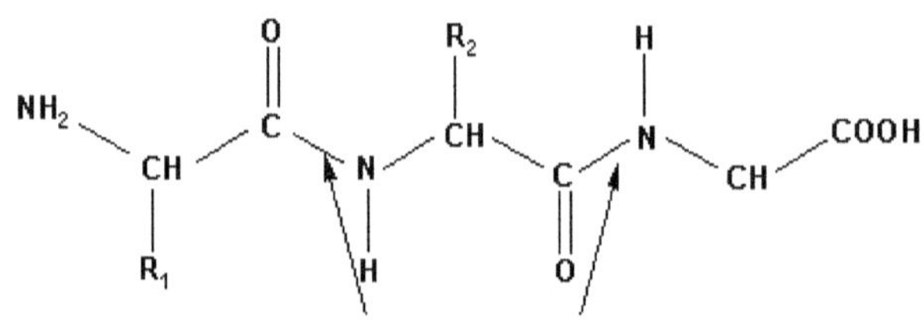

Two amino acids (dimer) with peptide bonds indicated by arrows

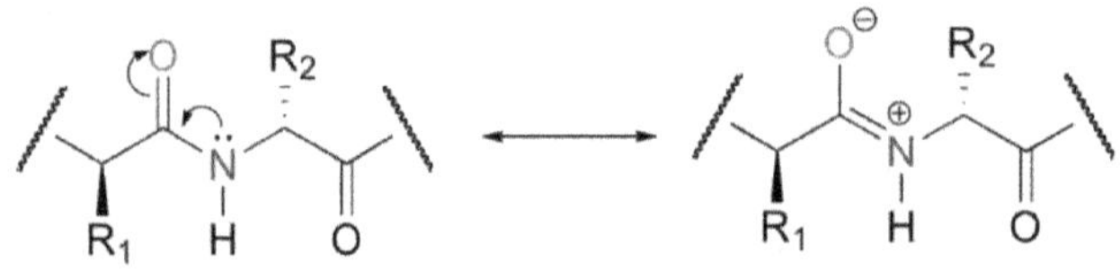

The peptide bond is rigid due to the resonance hybrids involving the lone pair of electrons on nitrogen, forming a double bond to the carbonyl carbon (and oxygen develops a negative formal charge).

8. C is correct.

Protonation or deprotonation of an amino acid residue changes its ionization state: it may become positively or negatively charged or neutral. The process may lead to changes in the interactions among amino acid side chains, as some ionic bonds may be compromised from the lack of opposite charge pairing.

Certain hydrogen bonding interactions may be modulated if Lewis bases are protonated with Brønsted acids, impairing their ability to accept hydrogen bonds from nearby amino acid residues.

9. D is correct.

Collagen is a protein that supports hair, nails, and skin.

Collagen is composed of a triple helix, and the most abundant amino acids in collagen include glycine, proline, alanine, and glutamic acid.

Much of the excess protein consumed in an animal's diet is used to synthesize collagen.

10. B is correct.

Secondary structure for proteins involves localized bonding.

The most important intermolecular interaction is hydrogen bonding, responsible for maintaining the alpha helix and beta pleated (parallel and antiparallel) sheet structures.

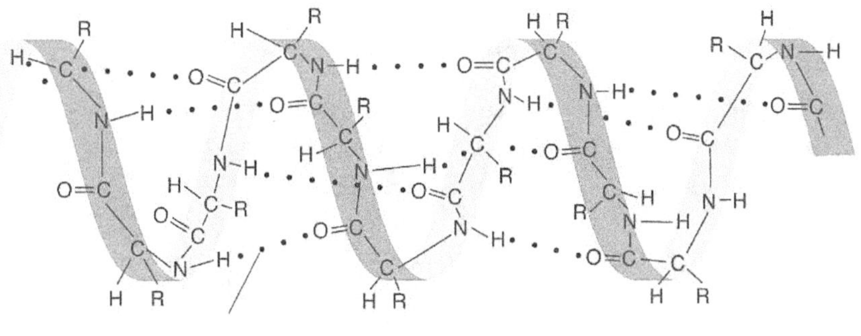

Alpha helix structure with hydrogen bonding shown as dotted lines

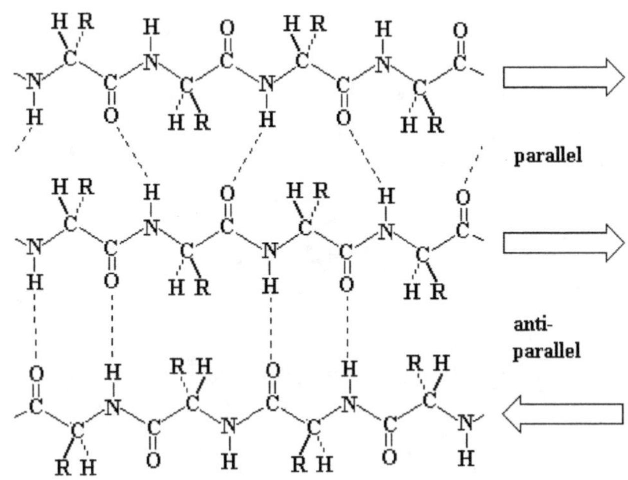

Beta pleated sheets (parallel and antiparallel) with hydrogen bonding shown

11. B is correct.

A *polypeptide chain* can undergo short-range bending and folding to form β sheets or α helices.

These structures arise as the peptide bonds can assume a partial double-bond character and adopt different conformations.

The arrangement of groups around the relatively rigid amide bond can cause *R* groups to alternate from side to side and interact with one another.

The carbonyl oxygen (C=O) in one region of the polypeptide chain could become hydrogen-bonded to the amide hydrogen in another region of the polypeptide chain. This interaction often results in forming beta-pleated sheets or an alpha helix.

Localized bending and folding of a polypeptide *do not* constitute a protein's primary structure.

continued...

A: *primary structure* of a protein is the amino acid sequence; individual amino acids are linked through peptide (i.e., amide) linkages.

C: *tertiary structure* is the 3-D shape that arises by further folding the polypeptide chain. Usually, these nonrandom folds give the protein a particular conformation and associated function.

D: *quaternary structure* is the spatial arrangement between two or more associated polypeptide chains (often linked by disulfide bridges between cysteine residues).

E: *zymogen* (or *proenzyme*) structure refers to an inactive enzyme that requires modification (i.e., hydrolysis to reveal the active site or change in configuration) to become an active enzyme.

12. B is correct.

Proteins are biological macromolecules composed of bonded amino acids with peptide (amide) bonds.

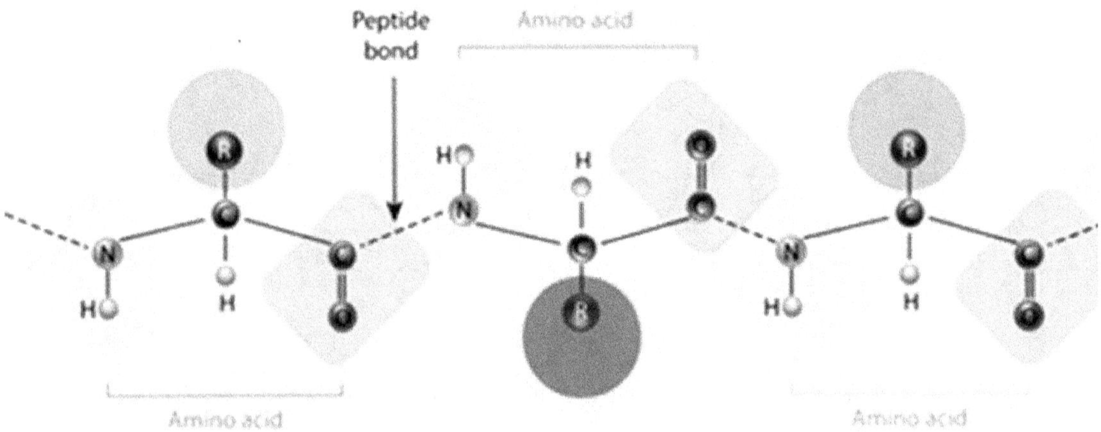

Three amino acids residues of nascent (i.e., growing) polypeptide

13. D is correct.

The plasma membrane is made of phospholipids (i.e., type of lipid). These molecules mostly possess nonpolar characteristics due to the long hydrocarbon chains, making the membrane permeable to nonpolar materials and semipermeable to polar or charged molecules.

14. B is correct.

Although cholesterol plays a pivotal role in the synthesis of other steroids and the integrity of cell membranes, too much cholesterol in the blood results in plaque deposits in blood vessels.

5. D is correct.

Steroid molecules are one of the two kinds of fat molecules, which are composed of fused rings.

Triacylglycerides are composed of long hydrocarbon chains with functionalized head groups.

16. C is correct.

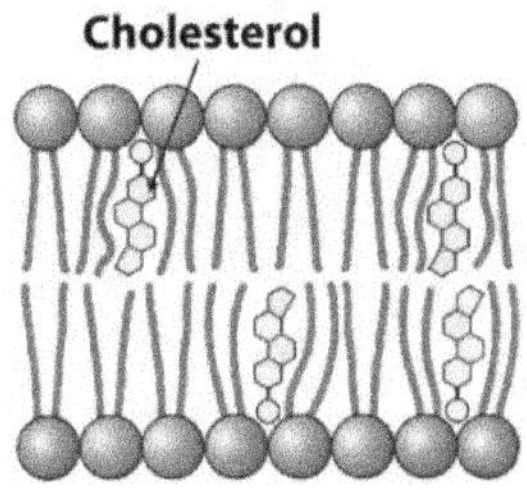

Cholesterol is a lipid molecule known as a steroid compound.

The fused ring structure of steroid molecules makes them rigid and has fewer degrees of motion due to the few conformations available for cyclic molecules *vs.* acyclic molecules.

Molecules such as phospholipids lack fused-ring structures, and they exist as straight-chained molecules.

Therefore, cholesterol in the cell membrane acts as a bidirectional regulator of membrane fluidity: at high temperatures, it stabilizes the membrane and raises its melting point, whereas, at low temperatures, it intercalates between the phospholipids and prevents them from clustering and stiffening.

17. B is correct.

Triglycerides (or triacylglycerides) are used for storage and exist in the adipose tissue of animals.

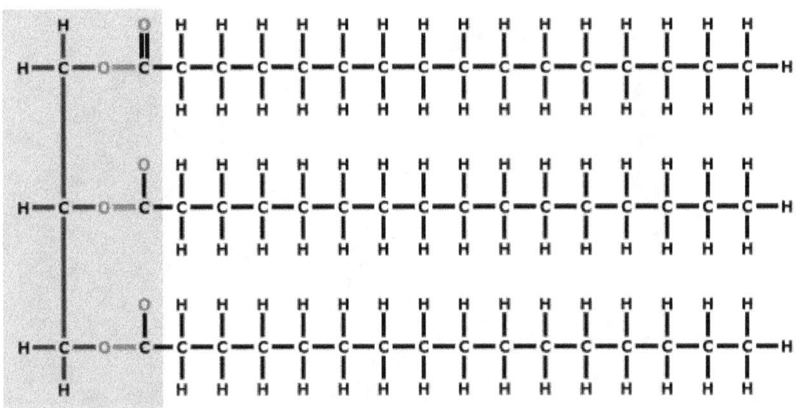

Triglyceride (glycerol and three saturated fatty acid chains)

Phospholipids(below) are the largest component of semi-permeable cell membranes.

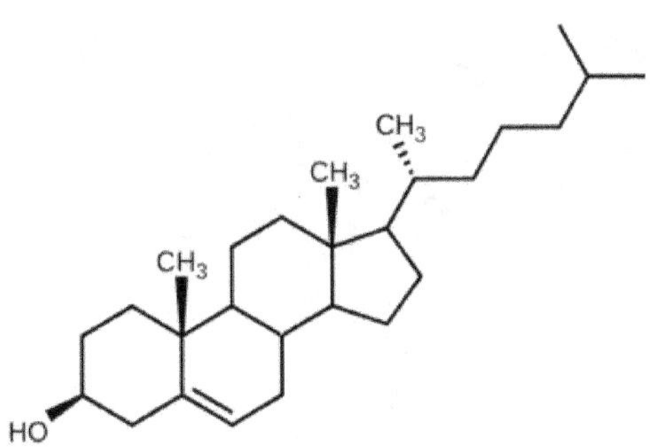

Cholesterol

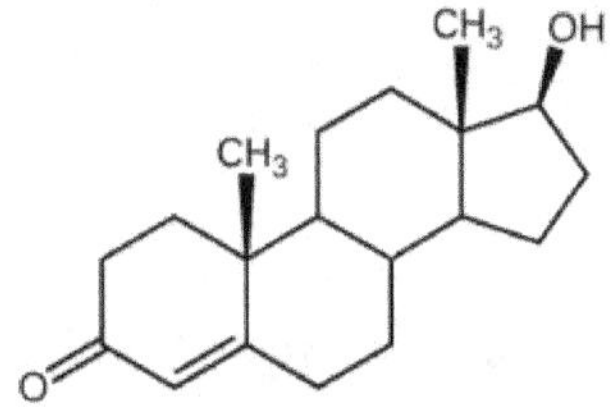

Testosterone

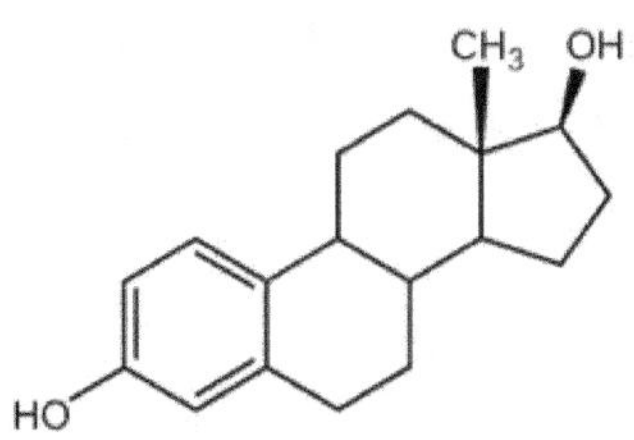

Estradiol

Cholesterol is the precursor molecule for several steroid hormones (e.g., progesterone, aldosterone).

Phospholipids differ mainly in the composition of the polar head region.

Steroids are lipids for cell signaling.

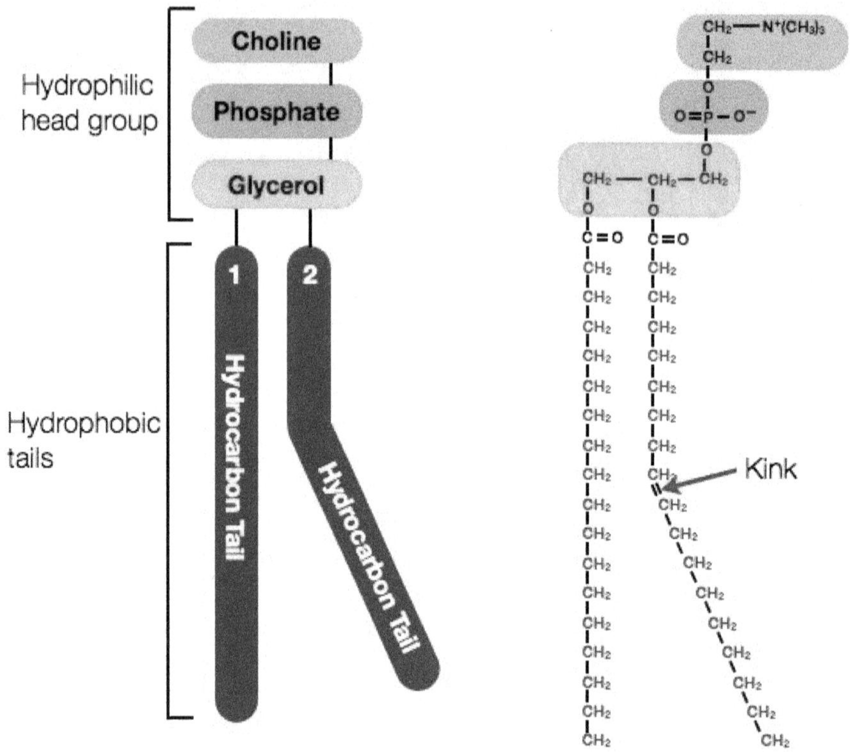

18. B is correct.

An omega-3 is when the alkene (E/Z) double bond is three carbon atoms from the methyl end.

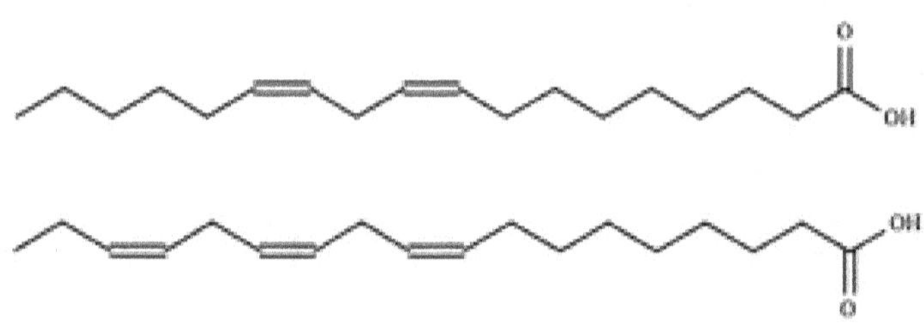

Linoleic acid (top)is ω-6,9 and linolenic acid (bottom) is ω-3,6,9. Note the positions of the double bonds and cis/trans (Z / E) relationships for double bonds

Linolenic acids are omega-3 fatty acids.

Linoleic acids are omega-6 fatty acids.

These molecules contain double bonds and are *unsaturated fats*.

19. D is correct.

Estradiol (shown) is a steroid hormone derived from cholesterol.

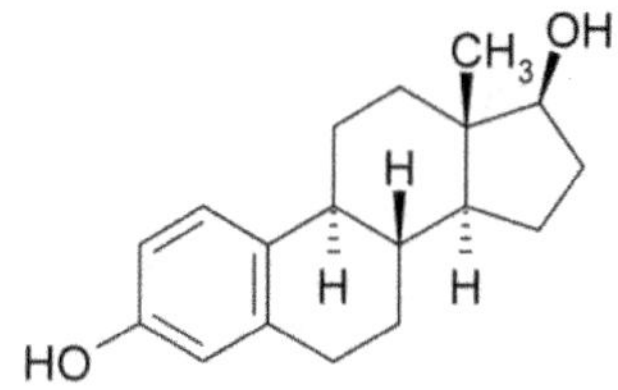

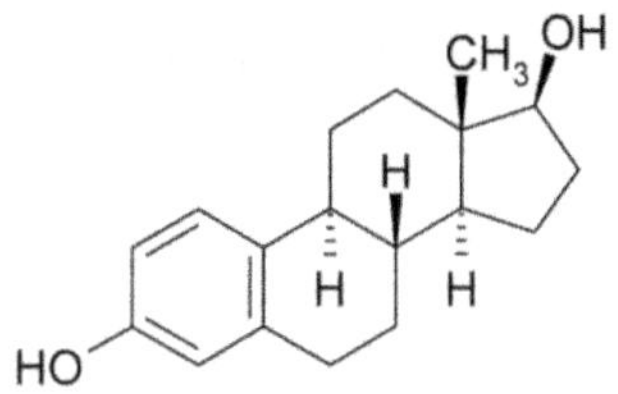

Estradiol is a derivative of cholesterol

Cholesterol is a lipid of four fused rings; three fused rings are six-membered, and the fourth is five-membered.

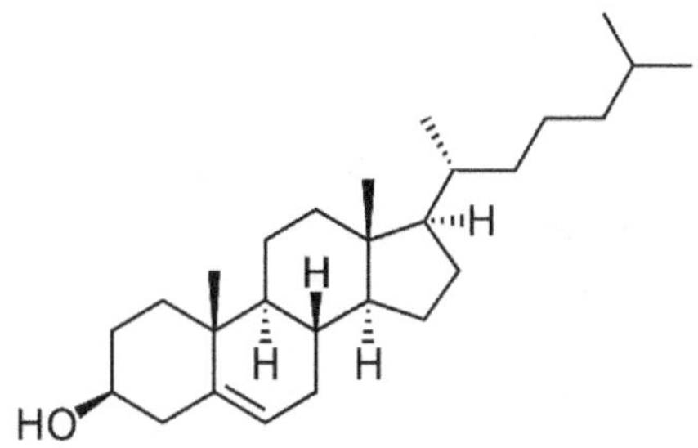

Cholesterol is a four-fused ring structure

Cholesterol is a steroid that makes up one of two types of lipid molecules.

Triglyceride (i.e., glycerol backbone with three fatty acid chains) is the other lipid (shown below).

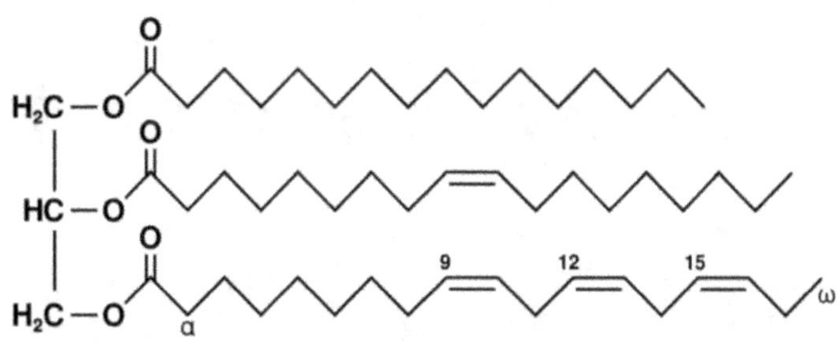

Triglyceride is glycerol backbone with three fatty acid chains

The carbon chain is numbered (example above) from the carboxyl end. Chemists number the double bonds as shown. Nutritionists specify the ω position from the terminal methyl group.

Position *omega* (ω) of the double bond(s) is the number of carbon atoms from the *terminal methyl group.*

20. B is correct.

Not all lipids are entirely hydrophobic. The ionic and polar heads of soaps and phospholipids, respectively, enable the molecules to interact with aqueous or polar environments.

The bulk of these molecules are hydrophobic because they largely consist of hydrocarbon chains or rings.

21. A is correct.

For unsaturated fats, the molecules are more likely to exist as oils (i.e., liquids) at room temperature.

Saturated fats tend to be solid at room temperature because the reduced forms of these molecules have better stacking properties, which allow them to form solid states.

22. B is correct.

The alkene molecules typically found in fatty acids tend to be *Z* alkenes.

Saturated Fatty Acid

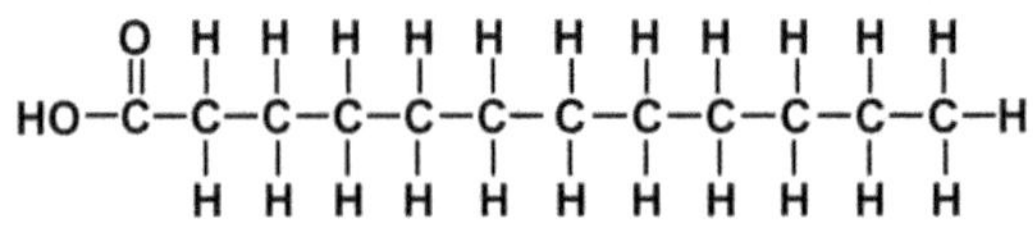

The double bond of the alkene (i.e., unsaturated) prevents the fatty acid molecules from stacking closely. The result is lowering of the melting point for the fat molecule. This may influence the state of matter of the oil, as unsaturated fats tend to be liquid at room temperature, and saturated fats tend to be solids.

Unsaturated Fatty Acid

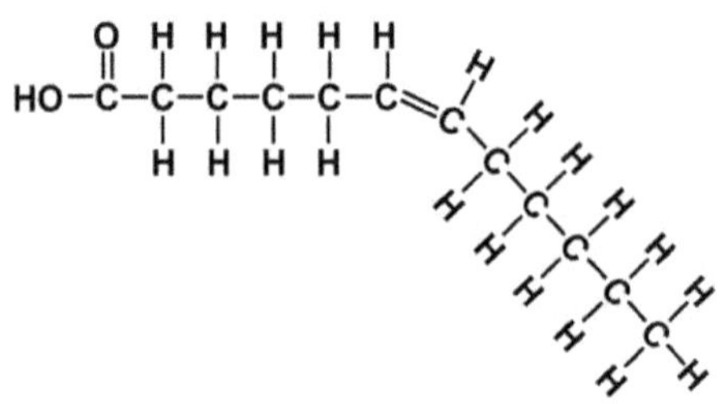

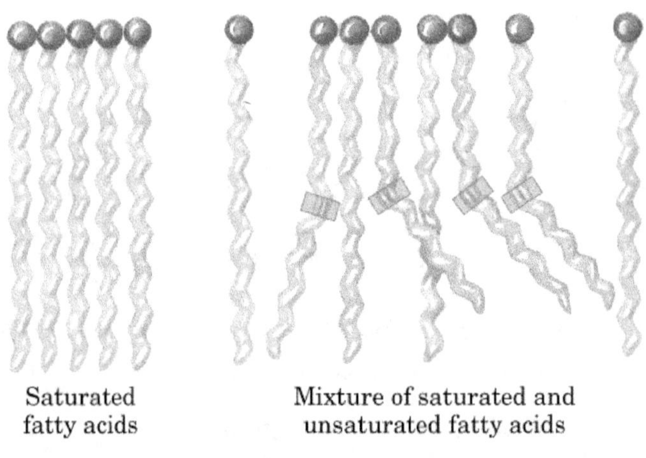

23. A is correct.

Oils are isolated from plant sources and may consist of several different fatty acids. These oils may contain saturated fat, but other fats present in the mixture are unsaturated fat molecules.

The saturated fats that may be present in oil are palmitic and stearic acid, and the unsaturated fats found in these oils include oleic, palmitoleic, and linoleic acids, as well as others.

24. B is correct.

There are two overall categories of lipids: long-chain lipids (e.g., triglycerides) and smaller, polycyclic lipids, such as steroids (e.g., cholesterol and its derivatives, such as estrogen and testosterone).

Lipid molecules are fat-soluble, and these molecules are largely soluble in organic/hydrophobic.

25. D is correct.

Saturated fats tend to be solid at room temperature because they lack alkene groups. The alkene groups in fat molecules lower the melting point for these compounds.

For example, butter is a dairy product made from the fat of cow's milk, and it is solid at room temperature and is mainly composed of saturated fat molecules.

26. D is correct.

Fatty acids are used to make fat molecules known as triglycerides.

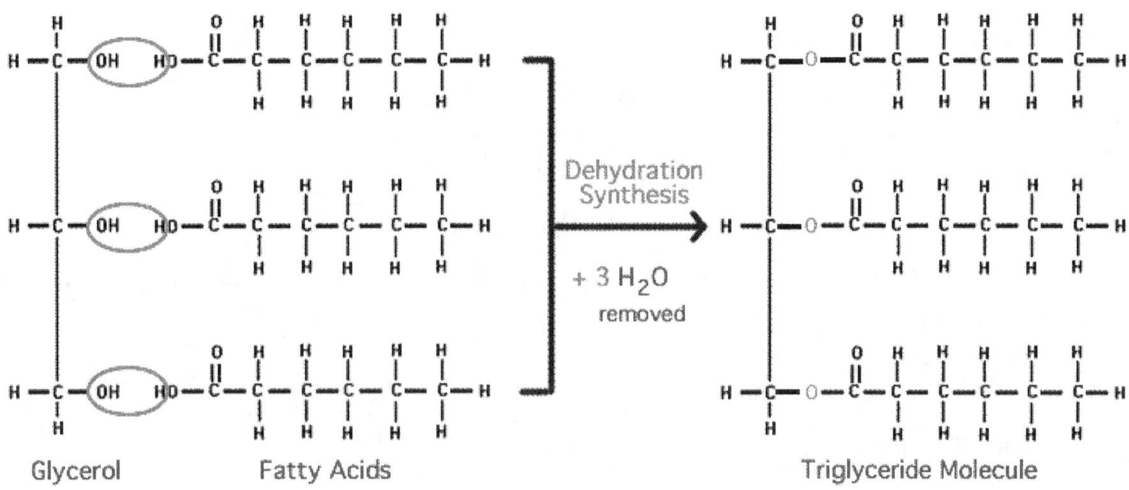

Formation, via dehydration (removal of H_2O), of triglyceride from glycerol and 3 fatty acids.

Fatty acids are made from one equivalent of a triol known as glycerol and three equivalents of acid-containing groups of fatty acids.

27. C is correct.

Phospholipids are important lipids that make up the bilayer structure of the membranes of cells, organelles, and other enclosed cellular structures.

Phospholipids are two (same or different) fatty acid molecules, a phosphate group, and a glycerol backbone.

The phospholipid contains a hydrophobic (i.e., fatty acid tail) region and hydrophilic (polar head) region.

The hydrophobic regions point toward each other in the membrane bilayer, while the polar heads point towards the inside (i.e., cytosolic) or outside (i.e., extracellular) sides of the bilayer.

28. D is correct.

Fat can be produced from glucose, but glucose is not produced from animal fat.

Excess glucose consumption can lead to increased levels of fat in the body.

29. C is correct.

One exception to the tendency for sugars to have oxygen atoms linked to every carbon is deoxyribose. It is similar to ribose; however, one of the alcohols is replaced with a carbon-hydrogen bond.

30. B is correct.

The ratio of these atoms that compose sugars is typically 1:2:1 (carbon : hydrogen : oxygen).

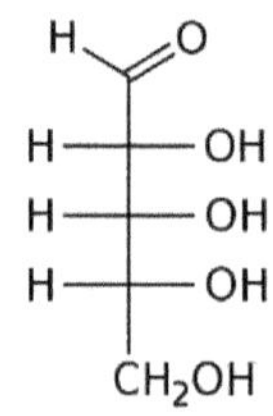

D-ribose: Fisher projections of an example ribose sugar

31. A is correct.

Monosaccharides cannot be broken down into simpler sugar subunits.

However, monosaccharides can undergo *oxidative degradation* to produce carbon dioxide and carbon monoxide when treated with nitric acid.

32. E is correct.

Lactose is a disaccharide composed of glucose and galactose.

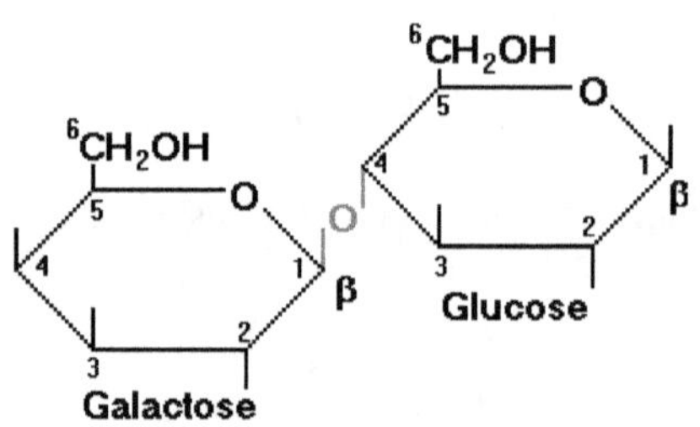

The glycosidic linkage in lactose is a β(1→4) linkage.

33. C is correct.

Glycogen is a polymer of glucose that functions as the energy store of carbohydrates in animal cells (plant cells use starch). Glycogen is common in the liver, muscle, and red blood cells.

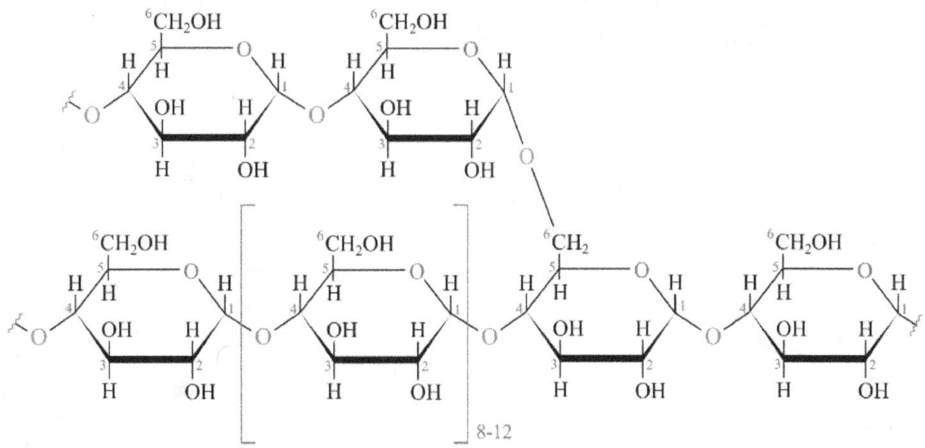

Glycogen is a large biomolecule consisting of repeating glucose subunits.

34. D is correct.

Disaccharides contain a glycosidic linkage that is an ether group. The ether can be protonated with Brønsted acids and hydrolyzed in the presence of water.

All polysaccharides can be hydrolyzed to produce monosaccharides (i.e., individual monomers of the polymer).

35. D is correct.

Monosaccharides are the basic unit of carbohydrates.

Subjecting these compounds to acids or bases will not hydrolyze them further. However, they can undergo oxidative decomposition by treating them with periodic acid to form formaldehyde and formic acid.

36. D is correct.

Carbohydrates can be more specifically described as organic compounds that contain carbon, hydrogen, and oxygen. The general molecular formula may vary depending on the type of carbohydrates, but many examples have the formula of $C_nH_{2n}O_n$.

37. C is correct.

The "di" prefix in the name suggests that there are two smaller subunits.

Monosaccharides are linked through glycosidic (i.e., oxygen bonded to two ethers) functional groups.

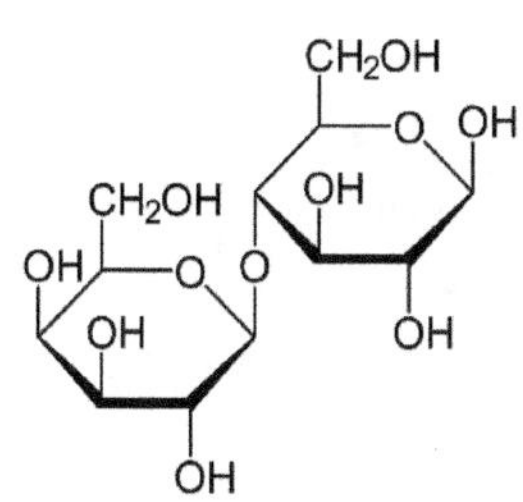

Lactose is a disaccharide formed by a β(1→4) linkage between galactose and glucose.

38. A is correct.

The *dipole interactions* that join the strands of DNA are specifically hydrogen bonds.

These bonds form from the acid protons between the amides and imide functional groups and the carbonyl and amide Lewis basic sites of the matched nitrogen base pairs.

39. D is correct.

The DNA molecule has a deoxyribose sugar-phosphate backbone with bases (A, C, G, T) projecting into the center to join the antiparallel strand of DNA (i.e., double helix).

The deoxyribose sugar-phosphate backbone is negatively charged due to the formal charge of the oxygen attached to the phosphate group.

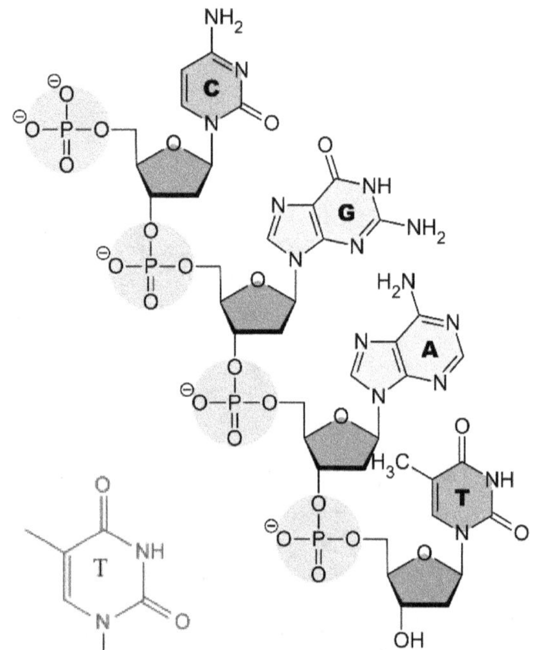

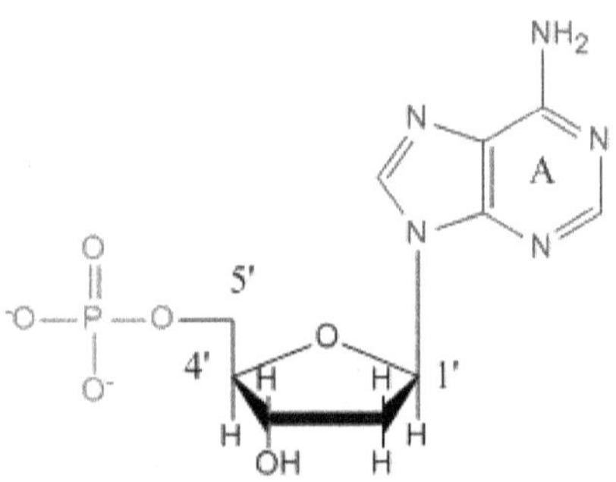

deoxyadenosine 5′-phosphate

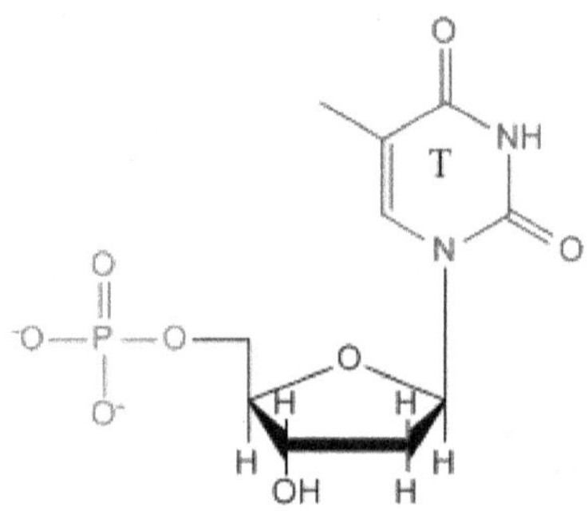

deoxythymidine 5′-phosphate

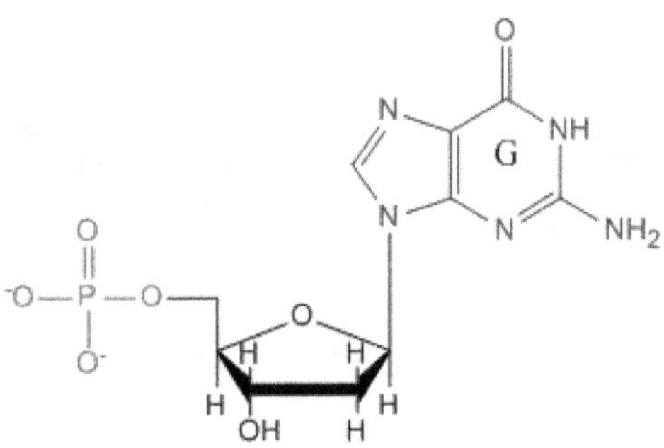

deoxyguanosine 5′-phosphate

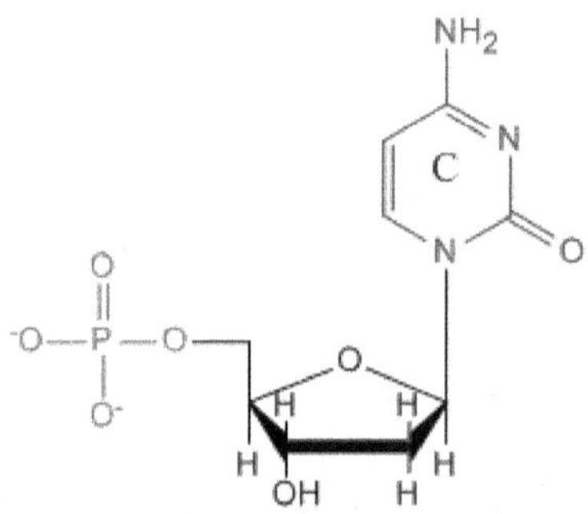

deoxycytosine 5′-phosphate

The purines (adenine and guanine) are double-ringed nitrogenous bases, while the pyrimidines (cytosine and thymine) are single-ringed nitrogenous bases.

A nucleotide has a deoxyribose sugar, base, and phosphate, while a nucleoside is a deoxyribose sugar and base without the phosphate group.

40. B is correct.

Four common nucleotides are in DNA molecules:

adenine (A), cytosine (C), guanine (G) and thymine (T)

Four common nucleotides are in RNA molecules:

adenine (A), cytosine (C), guanine (G) and uracil (U)

41. D is correct.

Because adenine forms two hydrogen bonds with thymine, and cytosine forms three hydrogen bonds with guanine. DNA polymerase (in the S phase of interphase) uses the complementary nitrogen base to synthesize the new (daughter) strand.

42. C is correct.

DNA molecules hold the genetic information of organisms.

RNA molecules are synthesized from the DNA strand to make proteins for the cell.

Genes are the sections of DNA responsible for the synthesis of proteins in cells.

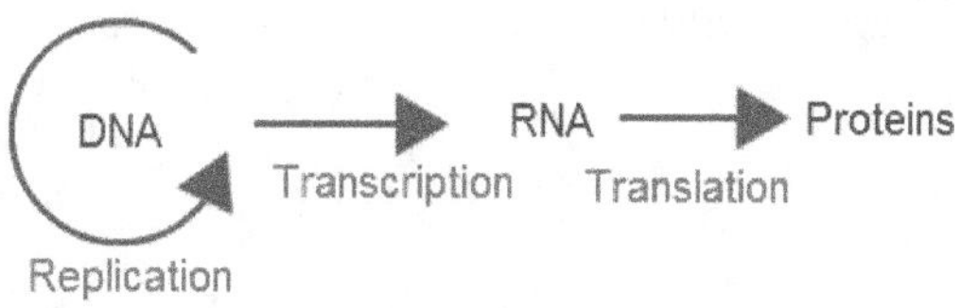

The central dogma of molecular biology (designates information flow)

43. D is correct.

There are five nucleotides, four of which appear in DNA.

These nucleotides are cytosine, guanine, adenine, and thymine.

In RNA molecules, the thymine is replaced with another pyrimidine nucleotide known as uracil.

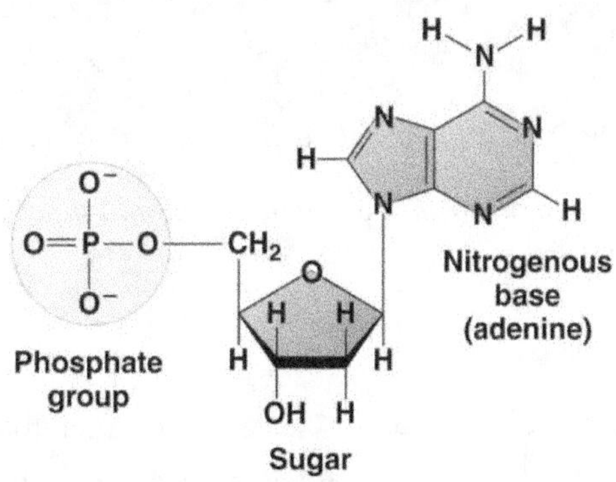

Nucleotides consist of a phosphate group (note the negative charge on oxygens), deoxyribose sugar (lack a 2'-hydroxyl), and a nitrogenous base (adenine, cytosine, guanine, or thymine)

44. D is correct.

DNA	DNA	mRNA	tRNA
A	T	A	U
C	G	C	G
G	C	G	C
T	A	U	A

Complementary base pairing for nucleotides

DNA → DNA (replication); DNA → RNA (transcription); RNA → protein (translation)

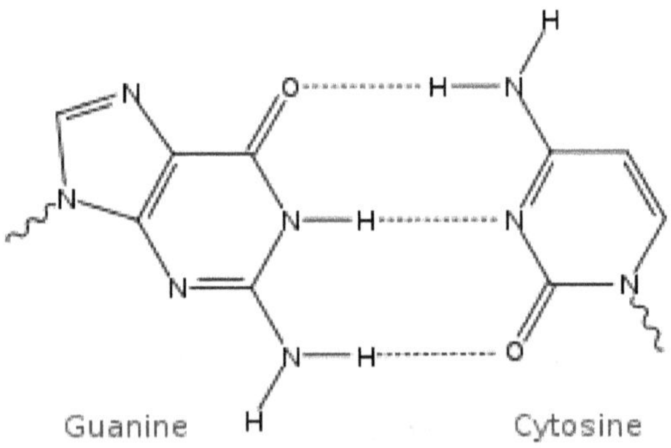

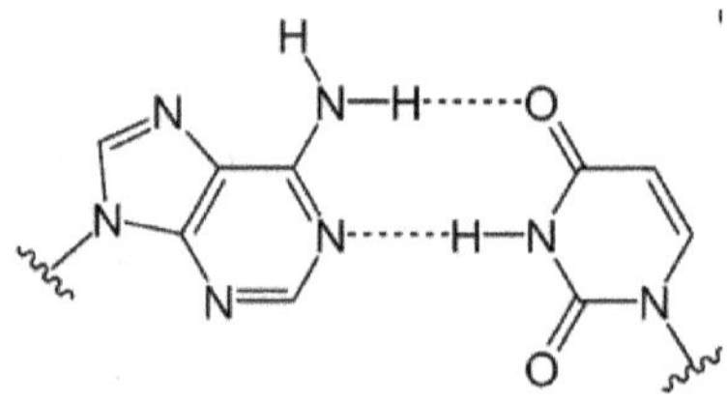

Adenosine (left) and Uracil (replaces thymine in RNA)

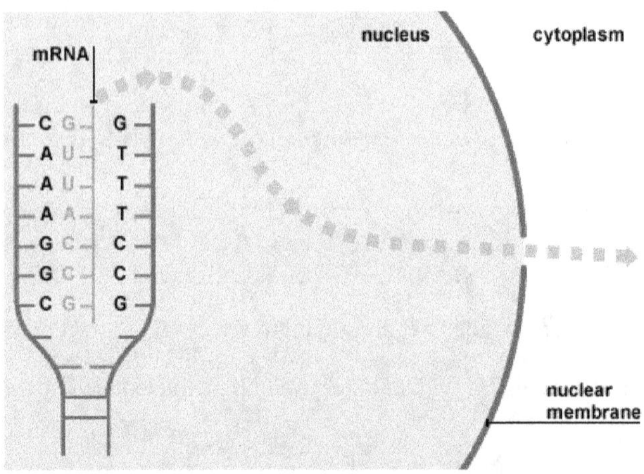

A single strand of DNA is the template for RNA synthesis during transcription.

The mRNA (after processing) is translocated to the cytoplasm for translation to proteins.

45. A is correct.

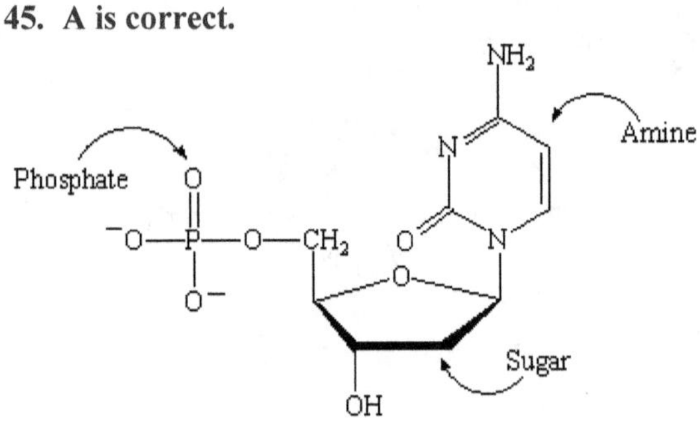

The hexose sugar, nitrogen base, and phosphoric acid group make up nucleotides, and these nucleotides are used to make larger molecules of nucleic acids.

Nucleotides consist of sugar, phosphate, and a nitrogenous base.

46. B is correct.

In any given molecule of DNA, each of the thymine nitrogen bases forms two hydrogen bonds with adenine.

Because each thymine pairs with an adenine residue, there are equal nitrogen bases in the molecule.

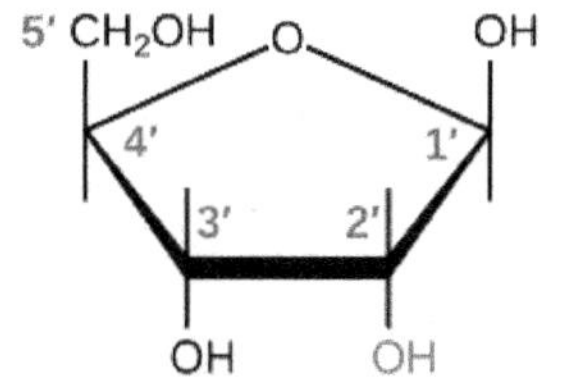

Ribose

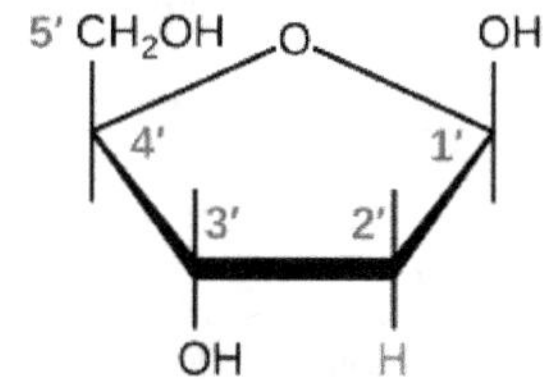

Deoxyribose

47. D is correct.

Ribose differs from deoxyribose sugars in that ribose lacks one alcohol (i.e., hydroxyl) group.

Ribose sugar (RNA) and deoxyribose sugar (DNA) are used to synthesize nucleic acid polymers.

48. A is correct.

Thymine is a pyrimidine nitrogenous base pair that forms two hydrogen bonds with adenine (purine) in the base-paired structure of DNA.

In RNA molecules, the nitrogenous base thymine is replaced by uracil.

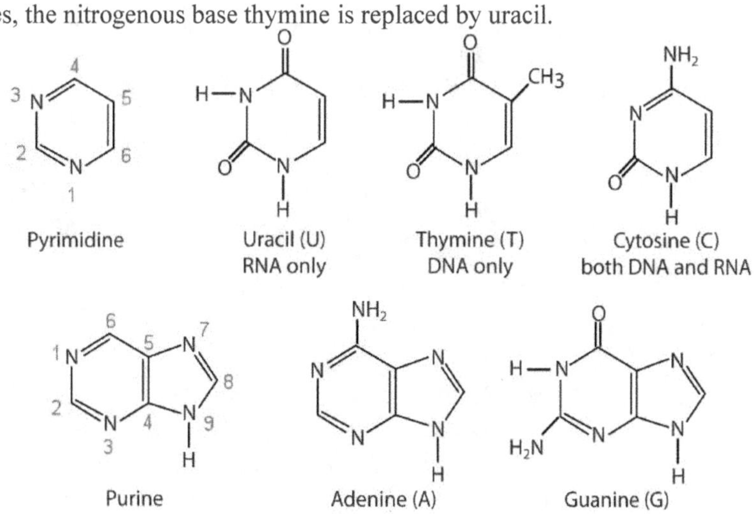

Purines (A, G) are single-ring structures, while pyrimidines (C, T, U) are double-ring structures.

49. B is correct.

There are three general components of nucleotides: a phosphate group, a cyclic five-carbon sugar, and a nitrogenous base.

Fat molecules are biomolecules for phospholipid membranes, storage fat molecules, and lipid molecules.

50. A is correct.

There are three major components of a nucleotide, the subunit that makes up nucleic acids.

All nucleic acids have a nitrogen base used for hydrogen bonding, a hexose sugar (ribose or deoxyribose), and a phosphate group that contains a phosphate linkage with the sugar.

Ester linkages are found in fats, glycosidic linkages are in sugars, and peptide linkages are in proteins.

51. C is correct.

Thymine is a pyrimidine nucleotide base that occurs in DNA but does not occur in RNA. Instead, RNA has uracil.

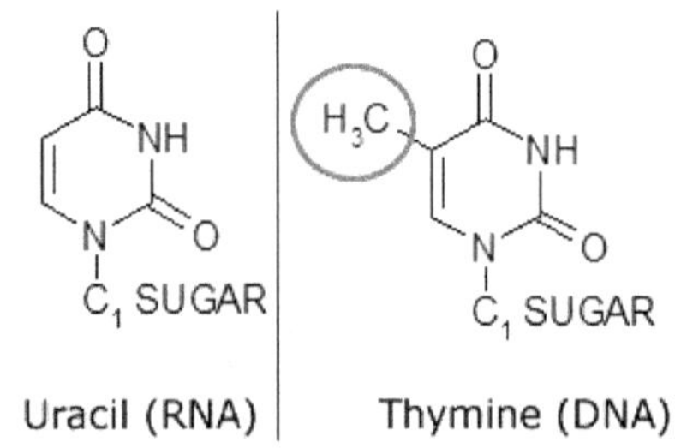

52. C is correct.

In DNA, thymine (T) hydrogen bonds with adenine (A).

However, in RNA, the thymine is exchanged for uracil (U).

53. C is correct.

Nucleic acids determine the sequences of amino acids because of groupings of nucleotides along a sequence corresponding to a particular amino acid.

The information of certain nucleic acids (i.e., mRNA) is translated on ribosomes with the help of tRNA.

Prions are infectious, disease-causing agents of misfolded proteins.

Notes for active learning

Notes for active learning

Specialized Cells and Tissues – Detailed Explanations

1. E is correct.

Ca^{2+} in muscle contractions binds the troponin-tropomyosin complex, exposing the myosin-binding sites.

2. A is correct.

Sodium (Na^+) has a positive charge, so the anion is chloride (Cl^-), which has a negative charge.

Potassium (K^+) is a cation with a positive charge.

Magnesium (Mg^{2+}) has a $^{2+}$ charge.

Lithium (Li^+) is a cation with a positive charge.

Calcium (Ca^{2+}) has a $^{2+}$ charge.

3. A is correct.

Caffeine does *not inhibit* the signaling pathway commonly stimulated by epinephrine.

*Use the graph to answer questions **4–8***

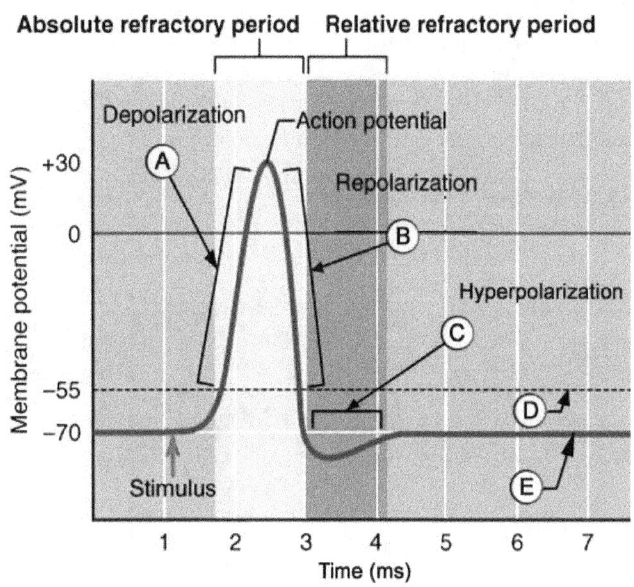

4. D is correct.

Arrow A points to *Na^+ ions entering axon.*

5. B is correct.

Arrow B points to *K^+ ions leaving axons.*

6. E is correct.

Arrow C points to the *slow close of K^+ voltage-gated channels.*

7. B is correct.

Arrow D points to the *threshold.*

8. D is correct.

Arrow E points to *resting potential.*

9. C is correct.

After acting upon the postsynaptic membrane, acetylcholinesterase inactivates acetylcholine (Ach) in the synaptic cleft.

Because chemical X deactivates acetylcholinesterase, ACh remains in the synaptic cleft and continues depolarizing the postsynaptic membrane.

10. D is correct.

Resting motor neurons are expected to exhibit a *more negative* resting potential than the threshold potential.

11. A is correct.

Na^+/K^+ ATPase transports 3 Na^+ ions *out* for every 2 K^+ ions *into* the cell. When the sodium/potassium pump is inhibited, intracellular $[Na^+]$ *increases* and intracellular $[K^+]$ *decreases*.

12. C is correct.

If a neuron's membrane potential goes from –70 mV to –90 mV, this is *hyperpolarization.*

13. E is correct.

Bone and *cartilage* are connective tissues that connect and support tissues and organs and are related in their lineage and activities. Connective tissue functions to *bind and support tissues*.

Connective tissue consists of *loose* and *dense connective tissue* (subdivided into dense regular and dense irregular).

Special connective tissue consists of *reticular connective tissue*, *adipose tissue*, *blood*, *bone*, and *cartilage*.

14. C is correct.

Resting membrane potential (i.e., voltage results from differences in charge) across a nerve cell membrane depends on the unequal distribution of Na^+ and K^+ ions.

Na^+/K^+ pump is an active transport protein that maintains an electrochemical gradient across the membrane by pumping three Na^+ out and two K^+ into the cell.

Unequal pumping of ions causes:

1) cells to be *negative inside* relative to outside;

2) high $[Na^+]$ *outside* the cell relative to inside; and

3) high $[K^+]$ *inside* relative to outside.

The membrane is more permeable to K^+ than Na^+ ions, and the balance between the pump and the *leaky* membrane determines the resting potential (i.e., –70mV).

15. E is correct.

The first event when a resting axon reaches threshold potential is the *opening of* Na^+ *gates*.

16. D is correct.

Schwann cells (in the peripheral nervous system) and *oligodendrocytes* (central nervous system) synthesize myelin, which functions as insulation around the axon.

Myelinated axons transmit nerve signals *much faster* than unmyelinated ones.

Multiple sclerosis is a *demyelinating disease*, which manifests in highly myelinated, fast-conduction neurons.

17. D is correct.

Synaptic summation measures the postsynaptic response after stimulation by the presynaptic cell.

18. B is correct.

A receptor is required for a cell to respond to a signaling molecule.

19. A is correct.

Muscle contraction occurs when *actin thin filaments* and *myosin thick filaments* slide past each other to shorten the muscle and produce a contraction.

Neither of the filaments shortens nor elongates.

Mmuscle decreases in size because the filaments slide past each other to cause muscle contraction.

20. B is correct.

When an organism dies, its muscles remain in the contracted state of *rigor mortis* for a brief period.

The lack of ATP to break bonds between thick and thin filaments is primarily responsible for this phenomenon.

21. B is correct.

Na^+/K^+ ATPase pumps three Na^+ ions out for every two K^+ ions into the cell.

If activity is blocked, there is a net increase of positive charge in the cell, and the inside becomes less negative.

As the inside becomes more positive, it reaches threshold (i.e., about –50mV), and depolarization ensues.

This change in ion concentrations decreases the resting potential (i.e., –70mV).

Ouabain inhibits the Na^+/K^+ pump.

If the pump were not working, the consumption of ATP would decrease, $[Na^+]$ increases within the cell (the pump is not moving Na^+ out of the cell), while $[K^+]$ increases in the extracellular environment (the pump is not moving K^+ into the cell).

22. B is correct.

Dendrites are an extension of a nerve cell that receives electrochemical impulses from cells and carry them inwards toward the cell body.

Axons are nerve fibers distinguished from dendrites in that they usually transmit signals rather than receive them.

Axon terminus is the inflated portion of the axon that releases neurotransmitters.

T-tubules are extensions of the cell membrane that pass-through muscle cells.

Nodes of Ranvier are gaps in the myelin sheath around an axon or nerve fiber.

23. E is correct.

Myelin sheath covers the axon and increases conduction velocity.

Saltatory conduction occurs by permitting membrane depolarization only at the *nodes of Ranvier*.

Myelin sheath comprises lipids deposited by *Schwann cells* for peripheral nerve cells and *oligodendrocytes* for central nerve cells.

24. D is correct.

Synaptic cleft (i.e., synapse) is the space between the axon terminus (end of one axon) and the abutting dendrite between neurons.

After stimulation, the presynaptic axon releases a neurotransmitter across this cleft, which diffuses and binds receptors on the postsynaptic dendrite of the next neuron.

Neurotransmitter is released by the presynaptic axon, received by the postsynaptic dendrite, and unidirectional.

Dendrite is where the input is received and is (often) near the cell body.

Axon process of the neuron projects away from the cell body and extends toward the axon terminus (where neurotransmitters are released).

Myelin sheath covers the axon and increases conduction velocity (i.e., permits saltatory conduction by preventing (via insulation) the passage of ions during depolarization except at the nodes of Ranvier).

Myelin sheath, composed of lipids, is deposited by Schwann cells for peripheral nerve cells and oligodendrocytes for central nerve cells.

Spinal nerve is a bundle of nerves that enter and exit the spinal cord.

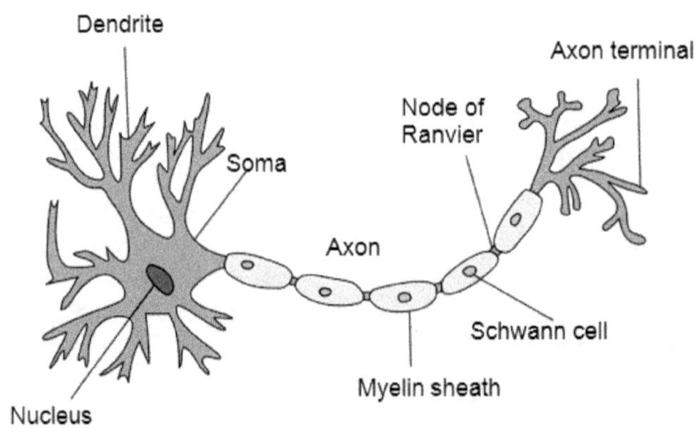

The dendrite receives input, the axon process propagates the all-or-none action potential, and the axon terminus releases neurotransmitters

25. A is correct.

In the communication link between a motor neuron and a skeletal muscle, the motor neuron is presynaptic, and the skeletal muscle is the postsynaptic cell.

26. B is correct.

Mitochondria self-replicate autonomously.

Mitochondria number in a cell varies widely by organism and tissue/type. Many cells have a single mitochondrion, while others contain several thousand mitochondria.

DNA is not an organelle.

Nucleus and *nucleolus* replicate and divide during mitosis and meiosis.

Ribosomes are not self-replicating.

Ribosomes consist of rRNA (synthesized in the nucleolus) and associated proteins; the cellular machinery assembles ribosomes.

27. C is correct.

Resting membrane potential of a neuron would be close to –70 mV.

28. C is correct.

In biochemistry (and pharmacology), a *ligand* is a molecule that forms a complex for a biological purpose.

Ligands include substrates, activators, inhibitors, and neurotransmitters.

In protein-ligand binding, a ligand usually is a signal-triggering molecule that binds a site on a target protein.

In DNA-ligand binding studies, a ligand is any molecule (e.g., protein or ion) binding to the DNA double helix.

Ligand binding to a receptor (ligand-receptor) alters a receptor protein's conformation (i.e., three-dimensional shape) that determines its functional state.

29. E is correct.

Myasthenia gravis causes the production of antibodies that remove acetylcholine (ACh) receptors from muscle fibers.

ACh cannot bind to membranes without these receptors and initiate depolarization.

None of the subsequent events occur without depolarization because the conduction of an action potential is impaired across the sarcolemma.

For muscle contraction, a neuromuscular junction is composed of a neuron's presynaptic membrane (i.e., axon terminus) and the postsynaptic membrane of a muscle fiber (i.e., sarcolemma).

Synapse is between two neurons (i.e., the space between is the synaptic cleft).

Presynaptic membrane of a neuromuscular junction releases the neurotransmitter acetylcholine (ACh) into the synapse from an action potential.

ACh diffuses across the synapse and binds acetylcholine receptors on the sarcolemma.

continued…

Binding of ACh causes the postsynaptic membrane to depolarize and generate an action potential, which causes the sarcoplasmic reticulum to release calcium ions into the sarcoplasm (i.e., the cytoplasm of a muscle fiber), which leads to the shortening of the sarcomeres and the resulting muscle contraction.

A: *acetylcholine* synthesis is not affected by *myasthenia gravis*.

Neurons synthesize acetylcholine and are independent of acetylcholine receptors on postsynaptic membranes.

Acetylcholine is a neurotransmitter in the autonomic nervous system (ANS).

Acetylcholine acts on the peripheral nervous system (PNS) and central nervous system (CNS) and is the *only neurotransmitter* used in the *motor division of the somatic nervous system.*

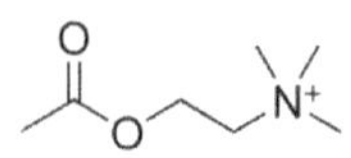

Acetylcholine

B: *muscle cells* store calcium in the *sarcoplasmic reticulum*, and Ca^{2+} release is affected by *myasthenia gravis*.

30. D is correct.

Thick myosin and *thin actin filaments* are in skeletal, cardiac, and smooth muscle.

31. E is correct.

Na^+ and K^+ are the major *ions* involved in the action potential of a neuron.

Blocking Na^+ channels and preventing Na^+ from entering the cell inhibits the propagation of the action potential because the movement of Na^+ creates depolarization.

–70mV (resting potential) becomes –40mV (threshold) for action potential propagation before the K^+ channels open to repolarize the neuron.

32. B is correct.

Glial cell is a non-neuronal cell that maintains homeostasis, forms myelin, and provides support and protection for neurons in the brain and peripheral nervous system

Oligodendrocyte is a glial cell that myelinates the axon of neurons in the CNS.

A: *astrocytes* are star-shaped glial cells in the brain and spinal cord and are the most abundant human brain cell.

Astrocytes perform many functions, including biochemical support of endothelial cells that form the blood-brain barrier, facilitate nutrients to the nervous tissue, and maintain extracellular ion balance.

Following traumatic injuries, *astrocytes* enlarge and proliferate to form a scar and produce inhibitory molecules that inhibit the regrowth of a damaged or severed axon.

C: *Schwann cells* myelinate axons of neurons in the PNS.

D: *choroid plexus* is a plexus (i.e., branching network of vessels or nerves) in the ventricles of the brain where *cerebrospinal fluid* (CSF) is produced.

Each of the four ventricles within the brain has a choroid plexus.

E: *chondrocytes* are the only cells in healthy cartilage and produce and maintain the cartilaginous matrix, which consists mainly of collagen and proteoglycans.

33. D is correct.

Neurons do not produce *myelin sheaths* but are synthesized by *Schwann cells* for peripheral nerve cells (PNS) and *oligodendrocytes* for central nerve (CNS) cells.

Schwann cells and *oligodendrocytes* wrap the axon to create layers of insulating myelin (composed of lipids), preventing the passage of ions along the axon process and increasing the rate of depolarization along the axon.

A: *axon hillock* is a region in the cell body (i.e., soma) where membrane potentials propagated from synaptic inputs (temporal and spatial summation) are aggregated (i.e., summed) before being transmitted to the axon.

In its resting state, a neuron is polarized, with its inside at about –70 mV relative to the outside.

Axon hillock "sums" depolarization events to determine if a sufficient magnitude of depolarization is achieved.

Once threshold is reached (between –50 and –40 mV), *all or none* action potential is propagated along the axon.

C: *nodes of Ranvier* are openings along the axons that permit depolarization of the membrane and give rise to saltatory conduction to increase the transmission rate along the nerve fiber.

E: *oligodendrocytes* are glial cells that deposit myelin around axons in the central nervous system, as do Schwann cells in the peripheral nervous system.

34. E is correct.

Resting muscle is not entirely relaxed but experiences a slight contraction known as tonus.

A: *tetanus* is a condition of sustained contraction due to an overlap of twitch impulses

B: *tonus* is the partial sustained contraction in relaxed muscles.

C: *isometric contractions* involve a constant length and increased muscle tension.

D: *isotonic contractions* involve the shortening of the muscle while the tension remains constant.

35. A is correct.

Organs are recognizable structures (e.g., heart, lungs, liver) performing specific physiological functions.

36. C is correct.

Na^+/K^+-ATPase pump moves K^+ *into* cells.

K^+ is *positively charged,* which causes the inside of the membrane to be more positive.

Resting potential (i.e., –70mV) of the neuron is measured with respect to the inside.

37. B is correct.

Na^+ / K^+ ATPase maintains the asymmetric concentration gradient of Na^+ and K^+ across the membrane.

38. D is correct.

Insulin binds to the receptor on the extracellular surface of the cell.

Insulin receptor (IR) is a tyrosine kinase transmembrane receptor activated by insulin, IGF-I, and IGF-II.

Ligand binding (e.g., insulin) to the α-chains of the IR domain induces a conformational change within the receptor and autophosphorylation of tyrosine residues within the intracellular TK domain of the β-chain.

Tyrosine kinase (e.g., insulin) receptors mediate their activity by adding a phosphate group to tyrosine on specific cellular proteins.

39. A is correct.

During an action potential, *Na^+ flows in,* and *K^+ flows out.*

Na^+ flows into the cell (down its electrochemical gradient) during an action potential to depolarize (make the voltage less negative) the cell membrane.

K^+ flows *out* (down electrochemical gradient) to repolarize the cell and restore the resting potential of –70mV.

When this outflow is excessive, the membrane becomes hyperpolarized due to the slow closing of K+ channels (e.g., the temporary value of –90mV), known as overshoot.

40. C is correct.

Folic acid (or vitamin B_9) is a water-soluble vitamin.

Folic acid is not biologically active, but its derivatives have high biological importance.

Folate is a naturally occurring vitamin in food, while folic acid is synthetically produced.

Folic acid (or folate) is essential for numerous bodily functions.

Humans cannot synthesize folate *de novo*, so it must be supplied through the diet.

Folate is needed to *synthesize, repair, and methylate DNA* and function as a cofactor in specific biological reactions. It is vital in rapid cell division and growth (e.g., infancy and pregnancy).

Children and adults require folic acid to produce healthy *red blood cells* and prevent anemia.

A: vitamin B_6 is a water-soluble vitamin.

Pyridoxal phosphate (PLP) is the active form and a cofactor in many reactions of amino acid metabolism (e.g., transamination, deamination, and decarboxylation).

PLP is necessary for the enzymatic reaction to release glucose from glycogen.

B: *calcium* is an essential component of a healthy diet and a mineral necessary for life.

Approximately 99 % of the Ca^{2+} is stored in bones and teeth. It has important uses, such as exocytosis, neurotransmitter release, and muscle contraction.

In heart muscles, Ca^{2+} replaces Na^+ as the mineral that depolarizes the cell to proliferate action potentials.

continued…

D: vitamin B_{12} (or *cobalamin*) is a water-soluble vitamin with a critical role in the ordinary functioning of the brain and nervous system and the formation of blood.

Vitamin B_{12} is involved in the metabolism of every cell, primarily affecting DNA synthesis and regulation, fatty acid synthesis, and energy production.

Fungi, plants, and animals are not capable of producing vitamin B_{12}.

Bacteria and archaea produce it with enzymes required for its synthesis.

Many foods are a natural source of B_{12} because of bacterial symbiosis.

Vitamin B_{12} is the largest and most structurally complicated vitamin produced industrially only through bacterial fermentation synthesis.

E: vitamin B_7 (biotin) is a water-soluble vitamin coenzyme for carboxylase enzymes.

Vitamin B_7 is involved in gluconeogenesis, the synthesis of fatty acids, and amino acids isoleucine and valine.

41. D is correct.

Local anesthetics dissolve into the hydrophobic membrane and inhibit membrane-bound proteins.

Na^+ voltage-gated channel is essential for nerve conduction but is inhibited by the local anesthetics in a nonspecific process that blocks nerve conduction.

42. E is correct.

Enzymes are not a class of molecules used as neurotransmitters.

43. D is correct.

Cell surface receptors (i.e., transmembrane receptors) are specialized integral membrane proteins facilitating communication between the cell and the outside world.

Extracellular signaling molecules (e.g., hormones, neurotransmitters, cytokines, growth factors) attach to the extracellular receptor, which triggers changes in the function of the cell.

Signal transduction occurs when the binding of the ligand outside the cell initiates a chemical change on the intracellular side of the membrane.

44. A is correct.

Piloerection is when the hair stands on its end and is a sympathetic nervous system *fight-or-flight* response (i.e., parasympathetic is associated with *rest* and *digestion*).

Sensory neurons bring information *toward* the central nervous system from sensory receptors, while motor neurons carry signals *to effector cells from the central nervous system*.

For piloerection, a signal from the CNS is relayed to effector cells, which involves a motor neuron.

45. A is correct.

Cytoplasm includes the cytosol, a watery fluid inside the cell.

46. B is correct.

Ca^{2+} binds troponin C on the actin (thin) filament.

Ca^{2+} binds, and tropomyosin changes its conformation to expose the myosin (thick filament) binding sites on the actin filament.

When myosin binds the myosin-binding sites on actin, the myosin head undergoes a *power stroke,* and the actin-myosin filaments slide past each other as the muscle contracts.

Plasma Ca^{2+} levels are tightly regulated by *calcitonin* and *parathyroid hormone* (PTH).

PTH raises calcium levels by stimulating osteoclasts, while calcitonin lowers plasma calcium levels by stimulating osteoblasts.

47. C is correct.

Myelinated axons exhibit greater conduction velocity than non-myelinated axons.

Myelin sheath, composed of lipids, is deposited by Schwann cells for peripheral nerve cells and oligodendrocytes for central nerve cells.

Myelin sheath covers the axon and increases conduction velocity (i.e., permits saltatory conduction by preventing (via insulation) the passage of ions during depolarization, except at the nodes of Ranvier).

48. A is correct.

Axon hillock is a specialized part of the cell body (or soma) of a neuron that connects to the axon and where membrane potentials propagated from synaptic inputs are summated before being transmitted to the axon.

Axon hillock is the *accounting center*, where a graded potential is summed (spatial or temporal summation) to be sufficient to reach the threshold.

Inhibitory postsynaptic potentials (IPSPs) and *excitatory postsynaptic potentials* (EPSPs) are summed in the axon hillock.

Once a triggering threshold is exceeded, an action potential propagates – all or none – through the axon.

Glial cell is a non-neuronal cell in the central nervous system involved in depositing myelin around axons.

49. E is correct.

Myelin acts as an *insulator*, preventing ions from passing through the axon membrane and allowing axons to conduct impulses faster.

Ions can only permeate through *nodes of Ranvier*, which are channels as small gaps in the myelin sheath.

Action potentials *skip* (i.e., saltatory conduction) from node to node and are much faster than conduction through a non-myelinated neuron because the area requiring depolarization to permit Na^+ in / K^+ out is smaller.

A: *action potentials* are initiated by graded stimuli that cause depolarization of the axon hillock.

B: pumping of Na^+ *out* of the cell is achieved by the Na^+/K^+ pump.

C: *resting potential* is achieved by the Na^+/K^+ pump.

D: *voltage-regulated ion channels* determine the threshold of a neuron.

50. D is correct.

Epithelial tissue forms coverings, linings, and glands.

51. C is correct.

Nodes of Ranvier are unmyelinated regions along the axon process between Schwann cells, depositing myelin.

Myelin sheath is a lipid that increases propagation speed for the action potential to travel from the cell body to the axon terminal.

Electrical impulses move by depolarization of the plasma membrane at the unmyelinated (nodes of Ranvier) and "jump" from one node to the next (i.e., *saltatory conduction*).

Acetylcholine receptors are on the dendrite of a postsynaptic neuron.

52. A is correct.

Organs are usually made of several tissues and many types of cells.

53. C is correct.

Purkinje fibers in cardiac tissue have large diameters and short axons.

Conduction velocity is related to diameter and inversely related to length.

Large diameter axons have more volume for action potentials to depolarize membranes, and more ions migrate.

Increasing the *diameter* or decreasing *axon length* increases the conduction velocity.

Cardiac Purkinje cell conduction is fast and Na^+ channel-dependent because conduction is channel-dependent.

54. B is correct.

Four primary *tissue* types are *connective*, *muscle*, *nervous* and *epithelial.*

55. E is correct.

Release of neurotransmitters requires the influx of Ca^{2+} and occurs at the *axon terminus*.

A: Na^+ influx causes neuron depolarization.

B: K^+ efflux (not influx) determines the neuron's resting potential and may cause hyperpolarization.

C: Mg^{2+} ions are involved in important processes such as the modulation of ion channels, receptors, and cell excitability in the central nervous system.

D: Cl^- ions are inside neurons and contribute to the negative resting potential (i.e., –70mV).

56. D is correct.

Adipose tissue (along with cartilage, bone, and blood) is connective tissue, mainly beneath the skin and around internal organs.

In adipose tissue, glucose is transformed and stored as *fat*.

Fatty acids are an essential energy source, ideal for an organism to store fat in the adipose tissue when a large quantity of energy needs to be stored for later use (e.g., hibernating bears).

57. B is correct.

Cells use Na^+ channels to maintain osmotic gradients and prevent lysis.

Skeletal, cardiac, smooth muscle and nerve cells are Na^+-dependent excitable cells producing action potentials when *depolarized* by Na^+ entering the cell.

Endothelial cells line luminal structures (e.g., gastrointestinal tract, blood vessels, excretory system).

Endothelial cells are *non-excitable* because they do not undergo depolarization.

58. E is correct.

Simple squamous epithelium comprises the cells lining the air sacs in the lungs.

Notes for active learning

Notes for active learning

Microbiology – Detailed Explanations

1. E is correct.

The *major* distinction between prokaryotic and eukaryotic cells is that prokaryotic cells do not have a nucleus, but eukaryotic cells do.

2. B is correct.

Penicillin prevents bacterial peptidoglycan cell wall formation by covalently binding to a serine residue at the transpeptidase active site.

Binding the enzyme's active site is characteristic of an *irreversible competitive* inhibitor.

Irreversible binding with a covalent bond creates a stable and permanent attachment.

Competitive because it competes with the substrate for the active site.

Penicillin (β-lactam antibiotic) inhibits the formation of peptidoglycan cross-links in the *bacterial cell wall.*

Enzymes hydrolyzing the peptidoglycan cross-links continue to function, but those that form crosslinks do not; this weakens the bacterium cell wall, and osmotic pressure rises, causing cell death (cytolysis).

A: *reversible competitive inhibitors* utilize weak molecular attachment (i.e., van der Waals or hydrogen bonds) when attaching to the enzyme's active site.

C: penicillin does *not digest* the cell wall but prevents cross-linking bacterial *cell wall peptidoglycan.*

The effect inhibits the bacteria's ability to remodel their cell wall or prevent progeny from forming a necessary cell wall to counter osmotic pressure.

E: *noncompetitive inhibitors* bind to an allosteric site, which is different from the active site.

Binding of a molecule to the allosteric site of the enzyme causes a conformational change (i.e., three-dimensional shape change), affecting the shape (i.e., function) of the active site.

3. D is correct.

Viruses are simple, non-living organisms that take on living characteristics when they infect a host cell.

Their genetic material is RNA or DNA; not arranged into chromosomes but associated with a complex of nucleic acids and histone proteins.

Viruses contain DNA or RNA and a protein coat (i.e., capsule).

Cellular machinery and biomolecules in a host cell (prokaryotic or eukaryotic) are required for viral replication.

4. D is correct.

Replica plating technique of Joshua and Esther Lederberg demonstrated (in 1952) that streptomycin revealed the presence of streptomycin-resistant bacteria.

5. B is correct.

Operon is a functional unit of genomic DNA with a cluster of genes under the control of a single regulatory signal or promoter.

Genes (i.e., nucleotide sequences) are transcribed together into mRNA strands.

The genes contained in the operon are expressed together or not at all.

Several genes must be *co-transcribed* and *co-regulated* to be defined as an operon.

6. C is correct.

Neurospora is a fungus and haploid for most of its life cycle.

For fungus, a brief diploid (2N) stage after fertilization transitions via meiosis to produce haploid (1N) cells, which repeatedly divide via mitosis before entering another sexual cycle.

A: most fungi undergo meiosis and a sexual cycle.

B: separation of fertilization and meiosis is characteristic of the life cycle of plants.

D: *fertilization* immediately following meiosis is characteristic of diploid organisms.

7. D is correct.

Enzymes are biomolecules in prokaryotic cells.

8. A is correct.

Endomembrane system extends the nuclear envelope and includes endoplasmic reticulum and Golgi apparatus.

Golgi receives proteins within vesicles from the RER and modifies, sorts, and packages proteins destined for the secretory pathway (i.e., plasma membrane, exocytosis from the cell, or organelles within the cell).

Golgi is essential for synthesizing *proteoglycans* (i.e., components of connective tissue) in the extracellular matrix of animal cells.

Golgi primarily modifies proteins delivered from the rough endoplasmic reticulum.

It participates in lipid transport and the synthesis of lysosomes and is the site of carbohydrate synthesis.

B: *lysosomes* are organelles with low pH that function to digest intracellular molecules.

C: *peroxisomes* are organelles in most eukaryotic cells.

Peroxisomes function in the breakdown of very-long-chain fatty acids through *beta-oxidation.*

In animal cells, the very-long-chain fatty acids are converted to medium-chain fatty acids and subsequently shuttled to mitochondria.

They are degraded, via oxidation, into *carbon dioxide* and *water*.

continued...

D: *smooth endoplasmic reticulum* connects to the nuclear envelope and functions in metabolic processes (i.e., synthesis of lipids, phospholipids, and steroids).

It carries out the metabolism of carbohydrates, drug detoxification, attachment of receptors on cell membrane proteins, and steroid metabolism.

Cells secreting lipids, phospholipids, and steroids (e.g., testes, ovaries, skin oil glands) have a robust, smooth endoplasmic reticulum.

Smooth endoplasmic reticulum contains *glucose-6-phosphatase* (enzyme), which converts glucose-6-phosphate to glucose during *gluconeogenesis*.

9. C is correct.

Hfr (high-frequency recombination) cell is a bacterium with a conjugative plasmid (F factor) integrated into its genomic DNA instead of being in an autonomous circular DNA element in the cytoplasm (i.e., a plasmid).

F^+ denotes cells with the F plasmid, while F^- denotes cells that do not.

Unlike a typical F^+ cell, Hfr strains attempt to transfer their *entire* DNA through the mating bridge (pili).

The F factor tends to transfer during conjugation, and often, the entire bacterial genome is dragged along, but the transfer is often aborted before the complete plasmid is transferred.

Hfr cells are useful for studying gene linkage and recombination.

Because the genome's transfer rate through the mating bridge is constant, investigators can use the Hfr strain of bacteria to study genetic linkage and map the chromosome.

A: during conjugation, the transfer of Hfr DNA is interrupted by spontaneous breakage of the DNA molecule at random points.

F^+ denotes cells with the F plasmid, while F^- denotes cells that do not.

Typically, the chromosome is broken before the F factor is transferred to the F^- cell.

Therefore, the conjugation of a Hfr cell with an F^- cell does not usually result in an F^+ cell, and the F factor usually remains in the Hfr cell.

B: Hfr cells produce sex pili (i.e., mating bridge in this example), and the F^- cell is the transfer recipient.

D: F factor is integrated into the bacterium's chromosome, and the F factor is replicated along with the cells' genome before conjugation.

continued...

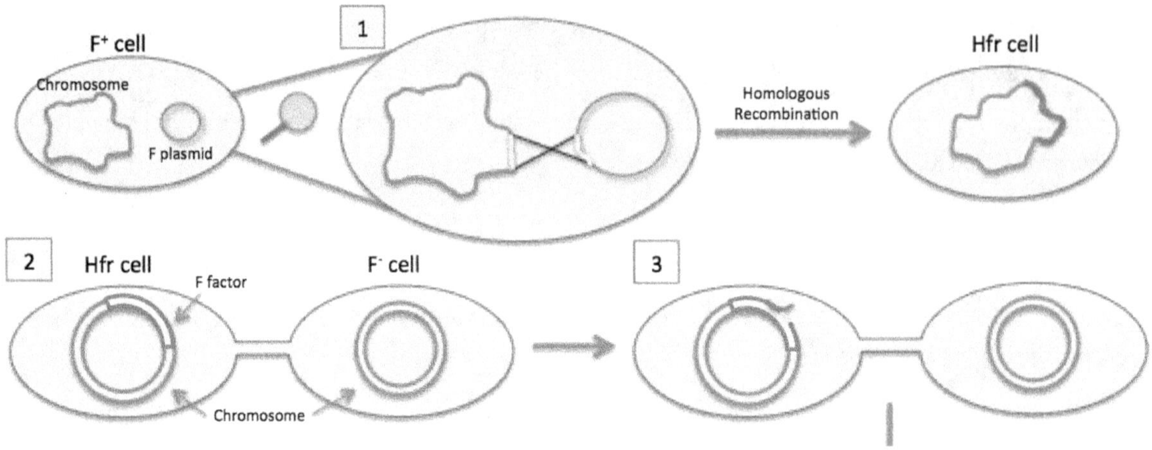

F^+ cell integrates plasmid to form Hfr cell that transfers the F factor by conjugation to an F^- cell

10. E is correct.

Ames test (developed in the 1970s) uses bacteria to evaluate whether a chemical likely causes cancer.

It assesses the mutagenic potential of chemical compounds.

Positive test indicates that the chemical is *mutagenic* and may function as a *carcinogen* because cancer is often linked to genetic mutations.

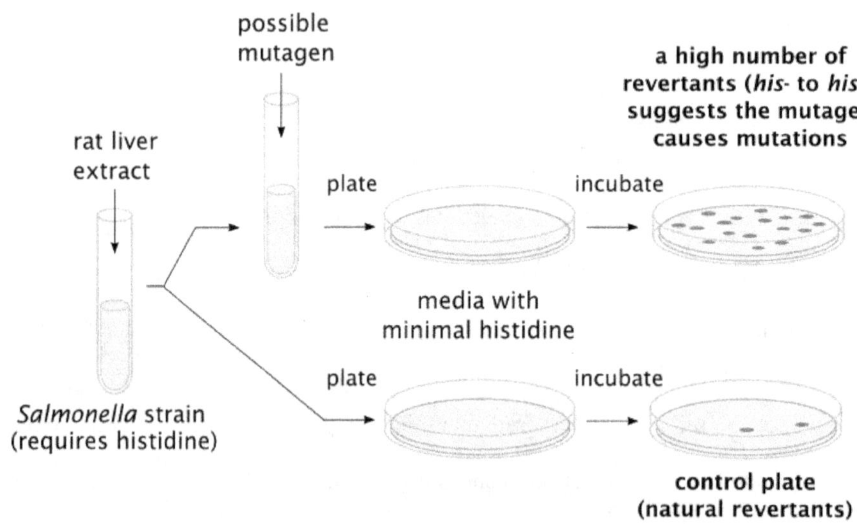

Ames test evaluates potential mutagens for the ability to transform nutrient-deficient cells upon exposure

11. A is correct.

Plants are generally autotrophs; animals are heterotrophs, and bacteria can be either.

Auxotrophs are unable to synthesize an organic compound required for growth.

In genetics, a strain is auxotrophic if it has a mutation that renders it unable to synthesize an essential compound (i.e., *essential* is needed for growth but cannot be synthesized).

continued...

For example, a uracil auxotroph is a yeast mutant with an inactivated uracil synthesis pathway gene.

Such a strain cannot synthesize uracil and will grow if uracil can be taken up from the environment.

B: *chemotrophs* obtain energy by oxidation of electron donors in their environments.

These molecules can be organic (e.g., chemoorganotrophs) or inorganic (e.g., chemolithotrophs).

Chemotroph designation contrasts with phototrophs, which utilize solar energy.

Chemotrophs can be autotrophic or heterotrophic.

C: *heterotrophs* use energy derived from another organism's metabolism (*hetero* means other).

D: *prototrophs* are characterized by synthesizing the compounds needed for growth.

12. B is correct.

Endospore is a dormant, tough, non-reproductive structure produced by certain bacteria.

Endospore consists of bacterium DNA and cytoplasm, surrounded by a tough outer coating.

Endospores can survive without nutrients.

Lack of nutrients usually triggers endospore formation.

13. E is correct.

Fungi are a large group of eukaryotes, including microorganisms such as mold, yeasts, and mushrooms.

Fungi are a kingdom (Fungi) separate from plants, animals, protists, and bacteria.

One significant difference is that fungal cells have cell walls with chitin.

The cell walls of plants and some protists contain cellulose, while bacterial cell walls contain peptidoglycan.

Fungi reproduce by sexual and asexual means and produce spores like basal plant groups (e.g., ferns, mosses).

Like algae and mosses, fungi typically have haploid (1N) nuclei with a small percentage of their life cycle in a diploid (2N) phase.

14. C is correct.

Cyanide is a poison interfering with the electron transport chain (ETC) of the inner plasma membrane (for prokaryotes) or the inner mitochondrial membrane (for eukaryotes) by binding to cytochromes (mainly *cytochrome c oxidase*) of the electron transfer complexes.

Cyanide inhibits the flow of electrons down their reduction potential and effectively inhibits the transport chain.

Consequently, the proton pump stops, and ATP cannot be generated aerobically.

Electron carriers such as NADH and $FADH_2$ cannot be oxidized and yield high-energy electrons to the electron transport chain because they are blocked.

NAD^+ and FAD are not regenerated, and aerobic *respiration ceases*.

continued...

DNA replication, RNA transcription, and protein synthesis require energy (ATP or GTP) and are processes necessary for host cell functions and viral replication.

If aerobic ATP formation is inhibited, there will be a deficiency of ATP for cell functions and viral replication.

Energy is required for a bacteriophage (a virus that infects bacteria) to replicate.

15. E is correct.

Transformation is the genetic alteration of a cell resulting from the direct uptake, incorporation, and expression of exogenous genetic material (exogenous DNA) from its surroundings and passes through the cell membrane.

Transformation introduces genetic material into nonbacterial cells, including animal and plant cells.

B: *conjugation* transfers genetic material (plasmid) between bacterial cells by direct cell-to-cell contact or a bridge-like connection between two cells (i.e., pili).

Conjugation is a mechanism of *horizontal gene transfer*, as are transformation and transduction.

The other gene transfer mechanisms do not involve cell-to-cell contact (conjugation uses a pilus).

C: *transduction* is how a virus transfers DNA from one bacterium to another.

In molecular biology, foreign DNA is introduced via a viral vector into another cell.

D: *recombination* is when DNA molecules exchange genetic information, forming new allele combinations.

In eukaryotes, genetic recombination between homologous chromosomes during prophase I of meiosis leads to a novel set of genetic information passed on to progeny.

Most recombination occurs spontaneously to increase genetic variation.

16. E is correct.

Viruses can have a genome consisting of double-stranded DNA (dsDNA).

Retroviruses have RNA genomes (single-stranded or double-stranded).

After infecting the host cell, the retrovirus uses *reverse transcriptase* to convert its RNA into DNA.

17. B is correct.

Cells of the same strain transfer their genomes in the same order and at the same rate.

Transfer is interrupted at times, and by matching the genes transferred to the length of time necessary for the transfer, the linear order of the genes is mapped.

A: *polycistronic* refers to the expression of bacterial genes on a single mRNA in an operon.

C: *rate of chromosome (i.e., F factor) transfer* must be constant because differences in the transfer rate would obscure the results.

D: F factors and the bacterial chromosome replicate using the same conjugation method (bacterial mating).

18. C is correct.

Prokaryotic ribosomes (30S small & 50S large subunit = 70S complete ribosome).

Eukaryotic ribosomes (40S small & 60S large subunit = 80S complete ribosome).

19. A is correct.

Retroviruses are a family of enveloped viruses that use reverse transcription to replicate within a host cell.

Retrovirus is a single-stranded RNA virus with nucleic acid as a single-stranded mRNA genome (including 5' cap and 3' poly-A tail).

Retrovirus is an obligate parasite within the host cell. Once inside the host cell cytoplasm, the virus uses its reverse transcriptase (enzyme) to produce DNA from its RNA genome. *Retro* describes this backward flow of genetic information.

Integrase (enzyme) incorporates this new reverse-transcribed DNA into the host cell genome. *Provirus* is the integrated retroviral DNA.

The host cell treats the viral DNA as part of its genome, translating and transcribing the viral genes and the cell's genes, producing the proteins required to assemble new copies of the virus.

20. C is correct.

Virus is a simple non-living organism that takes on living characteristics when it enters host cells.

Since viruses are not free-living organisms, they must replicate within a host cell.

Viruses contain DNA (single-stranded or double-stranded) or RNA (single-stranded or double-stranded) and have a protein coat.

Bacteriophage (i.e., a virus that infects bacteria) injects its genome into the bacterium while leaving the protein coat on the cell surface.

However, the entire virus (including the protein capsid) may enter the host cell in eukaryotes.

After entering the cytoplasm, the protein coat is removed (i.e., a virus is unencapsulated).

21. A is correct.

Bacterial conjugation transfers genetic material (plasmid) between bacterial cells by direct cell-to-cell contact or a bridge-like connection between two cells (i.e., pili).

Conjugation is *horizontal gene transfer*, as are transformation and transduction.

The other mechanisms do not involve cell-to-cell contact (conjugation requires a pilus).

B: *transformation* is the genetic alteration resulting from the direct uptake, incorporation, and expression of exogenous genetic material (exogenous DNA) from its surroundings and through the cell membranes.

In molecular biology, transformation introduces genetic material into nonbacterial cells, including animal and plant cells.

continued…

C: *transduction* is how a virus transfers DNA from one bacterium to another.

In molecular biology, *transduction* is when foreign DNA is introduced into another cell via a viral vector.

D: *recombination* is when DNA molecules exchange genetic information, resulting in a new combination of alleles.

Most recombination occurs spontaneously to increase genetic variation.

In eukaryotes, *genetic recombination* between *homologous chromosomes* during prophase I of meiosis leads to a novel set of genetic information passed on to progeny.

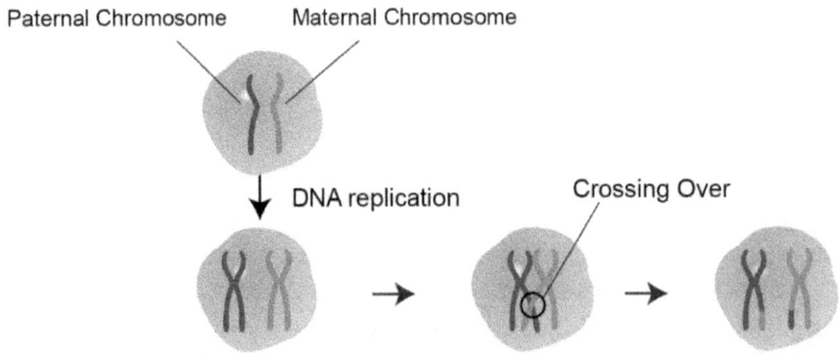

Crossing over occurs during prophase I at the tetrad during prophase I of meiosis as the product of recombination

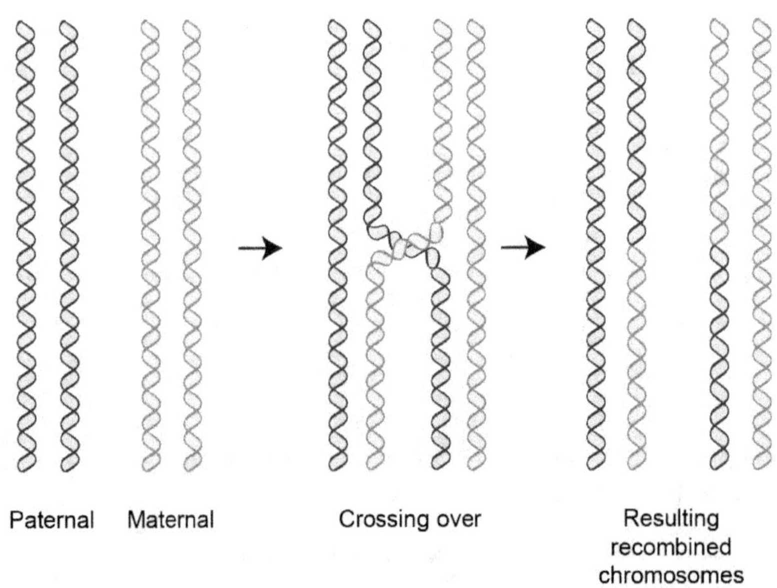

Recombination between homologous chromosomes during prophase I of meiosis

22. D is correct.

E. coli on the soft agar produces a solid growth (i.e., lawn) covering the agar.

Phages lyse cells, releasing phages (i.e., viruses).

Clear spots on the agar plate correspond to where the bacteria were lysed and are *plaques* (i.e., clear spots).

A: *viral replication* produces clear spots on the lawn of *E. coli* growth.

B: *colonies* on the agar surface indicate *E. coli* growth, not locations where the virus lysed the bacteria cells.

C: *smooth layer* of bacterial growth on the agar plate indicates the absence of the virus.

23. E is correct.

Gram staining differentiates bacterial species into Gram-positive and Gram-negative by the chemical and physical properties of peptidoglycan cell walls.

Gram stain is usually the first assay in identifying a bacterial organism.

Gram-negative bacteria are more resistant to antibiotics than Gram-positive bacteria, despite their thinner peptidoglycan layer.

Pathogenic capability of Gram-negative bacteria is often associated with specific components of their membrane, in particular, the lipopolysaccharide layer (LPS).

Penicillins are β-lactam antibiotics to treat bacterial infections by, usually, Gram-positive organisms.

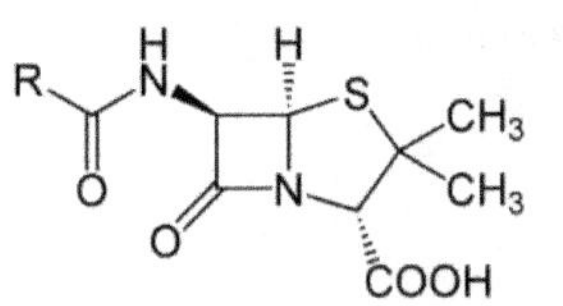

Penicillin core structure, where "R" is the variable group

Gram-positive bacteria take up the violet stain used in Gram staining.

This differential staining distinguishes them from the other large group of bacteria, Gram-negative bacteria that cannot retain the crystal violet stain.

Gram-negative bacteria take up the counterstain (e.g., safranin or fuchsine) and appear red or pink.

Gram-positive bacteria retain crystal violet stain by a thick *peptidoglycan layer* superficial to cell membranes.

In Gram-negative bacteria, this peptidoglycan layer is thinner and between two cell membranes.

The cell membrane, peptidoglycan layer, and cell wall are three distinct structures.

Cell walls provide structural support, protection, and rigidity to the cell.

Gram-positive bacteria have a *thicker peptidoglycan cell wall*, giving them a *purple-blue color* from staining.

Gram-negative bacteria stain a pink-red color.

continued…

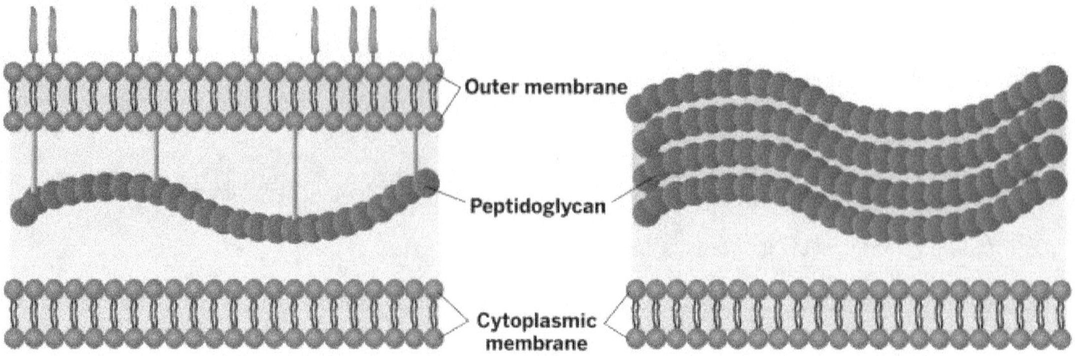

Gram-negative (left) comparison with gram-positive (right) cell walls

24. B is correct.

Viral genomes are single-stranded or double-stranded DNA, single-stranded or double-stranded RNA.

Retroviruses contain a single-stranded RNA (i.e., mRNA as its genome).

Retroviruses have *reverse transcriptase* and *integrase* to facilitate their infective ability within the host cell.

25. E is correct.

Operons are mainly in prokaryotes and encode for clusters of related (i.e., similar function) genes.

Lac operon of *E. coli* allows the digestion of lactose and consists of a set of control and structural genes.

Three structural genes are controlled by an operator on another part of the genome.

Lac genes encoding enzymes are *lacZ*, *lacY,* and *lacA*.

The fourth *lac* gene is *lacI*, encoding the lactose repressor—"I" stands for *inducibility*.

Without lactose, a repressor protein (*lacI*) is bound to the operator, preventing RNA polymerase from binding.

The binding of the repressor protein prevents the translation of the structural genes (*lacZ*, *lacY,* and *lacA*).

When lactose is present (i.e., glucose is absent), *lacI* binds the repressor dissociating from the operator region.

RNA polymerase (i.e., *inducible system*) attaches to the promoter, and translation occurs.

26. D is correct.

Prophage is a viral genome inserted and integrated into the circular bacterial DNA chromosome or as an extrachromosomal plasmid.

Lysogenic phase (i.e., latent form) of a bacteriophage is when viral genes are in bacterium without disrupting the cell.

continued…

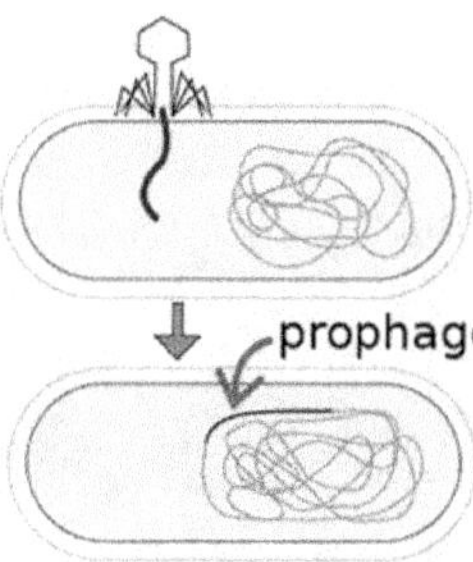

Prophase formation with a virus infecting a bacterial cell

Lytic cycle is one of the two cycles of viral reproduction, the other being the lysogenic cycle.

Lytic cycle destroys the infected cell and its membrane.

A critical difference between the lytic and lysogenic phage cycles is that in the *lytic phage*, the viral DNA exists as a separate molecule within the bacterial cell and *replicates separately from host bacterial DNA*.

The location of viral DNA in the lysogenic phage cycle is within the host DNA (i.e., integrated); the virus (phage) replicates using the host DNA machinery.

Phage is a free-floating separate molecule to the host DNA in the lytic phage cycle.

27. C is correct.

Retrovirus has an RNA genome transcribed (by the virus-encoded reverse transcriptase enzyme) into single- and then double-stranded DNA incorporated into the host cell genome.

B: *reverse transcriptase* (enzyme) converts the RNA of the virus into DNA when infecting host cells.

28. B is correct.

E. coli are normal flora of the human gut and are adapted to live in the human intestine.

Therefore, the optimal temperature for growth is 37 °C, whereby enzyme activity is optimal.

A: temperature of the human intestine is 37 °C. Bacteria growing optimally at this temperature exhibit the typical phenotype.

C: *conjugation* by bacteria is not necessary for growth.

Conjugation is mediated by a small genetic element (i.e., fertility or F factor) as independent or integrated into the bacterial chromosome.

F factor encodes for the F pili, which forms a conjugation bridge and allows genetic material transfer between mating cells.

Cells carrying the F factor are F^+ and transfer to an F^- cell.

Part of the bacterial chromosome can be transferred during conjugation, but the point of origin and the gene order are the same.

D: oxygen utilization is irrelevant to the growth temperature of bacteria.

29. A is correct.

Some bacteria can propel themselves through liquid using a *flagellum*.

30. C is correct.

When a virus infects a cell (eukaryotic or prokaryotic), it must attach to a host cell via specific proteins on the viral capsid or envelope. These proteins bind to receptor proteins on the membrane of the target cell.

Following this process, the virus enters cells via one of the three mechanisms: membrane fusion, endocytosis, or viral penetration. The latter mechanism is specific to bacteriophages.

Membrane fusion is the most well-known mechanism for enveloped viruses to enter eukaryotic cells.

The viral envelope fuses with the host cell's membrane, and the contents of the virus are released into the cell.

Endocytosis is utilized by enveloped and non-enveloped viruses that infect eukaryotic cells.

In endocytosis, the virus is engulfed by the cell and shuttled by cellular vesicles to the cytoplasm.

31. C is correct.

Bacteria are prokaryotes and have *no membrane-bound organelles* such as peroxisomes or nucleolus.

32. E is correct.

Lysogen is a bacterial cell where a phage (i.e., virus) exists as DNA in its dormant state (prophage).

Prophage is integrated into the host bacterial chromosome (i.e., lysogenic cycle) or (rarely) exists as a stable plasmid in the host cell.

The prophage expresses genes that repress the phage's lytic action until this repression is disrupted.

The virus enters the *lytic life cycle*, ultimately rupturing the cell and releasing virions.

B: *temperate* refers to the ability of some bacteriophages to enter the lysogenic lifecycle.

33. D is correct.

Protein synthesis (like for eukaryotes) occurs in three phases: initiation, elongation, and termination.

For translation, hydrogen bonds join codons on mRNA and anticodons on tRNA, not amino acid and mRNA.

The prokaryotic cellular translation machinery is the 70S ribosome (30S small and 50S large subunits).

F-met is used only in prokaryotic translation initiation.

First, an *initiation complex* forms with the 30S subunit, mRNA, initiation factors, and a special initiator formyl methionine F-met tRNA.

50S subunit binds the initiation complex to form the 70S ribosome.

The complete large subunit of the ribosome complex has two binding sites: the P (peptidyl transferase) site and the A (aminoacyl) site.

continued...

Initiator formyl methionine (F-met) tRNA is in the P site.

Elongation begins with binding a second tRNA (charged with its corresponding amino acid) to the vacant A site of the large ribosomal subunit.

The appropriate amino acid is determined by the complementary hydrogen bonding between the anticodon of the tRNA and the next codon of mRNA.

The orientation for the mRNA (from 5' to 3') is E-P-A (ribosomes move towards the 3' end of mRNA).

Peptide bonds are catalyzed by peptidyl transferase (at the P site) with a nucleophile attack from the lone pair of electrons on the N of the amino terminus on the carbonyl of the C terminus (at the A-site).

tRNA in the P site, now uncharged (without amino acid), moves to the E site and dissociates.

tRNA in the A site (after peptide bond formation and the ribosome translocates) is now in the P site, the A site is vacant, and this is where the incoming tRNA (carrying the next amino acid to join the polypeptide) binds.

This binding cycle to the A site, peptide bond formation, and translocation create a growing polypeptide chain.

Termination occurs when one of the three tRNA stop codons (i.e., not charged with an amino acid) hybridizes to the mRNA in the A site.

Termination factors catalyze the hydrolysis of the polypeptide chain from the tRNA, and the ribosomal complex dissociates.

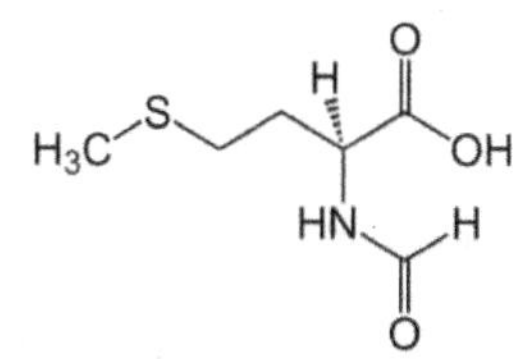

N-Formylmethionine (fMet) is a derivative of methionine whereby a formyl group adds to the amino group

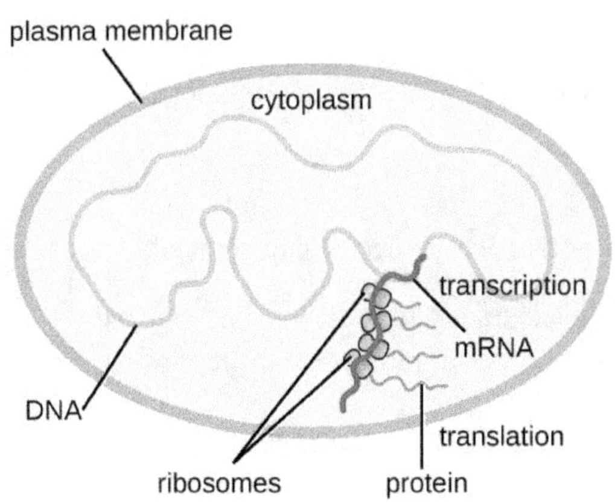

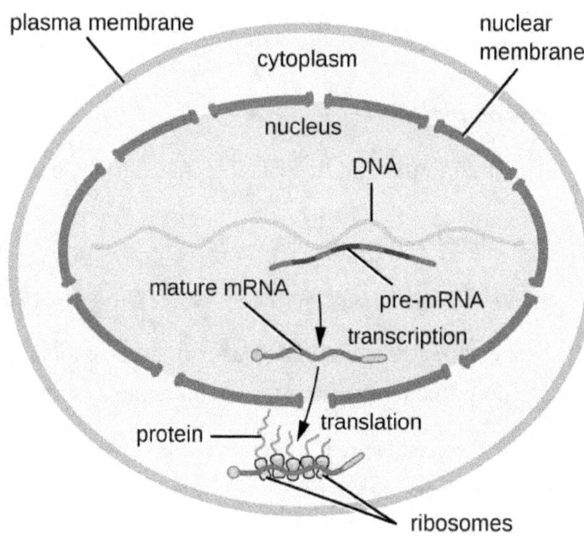

Prokaryotic cell (left) and eukaryotic cell (right). Prokaryotes have concurrent transcription and translation, while eukaryotes separate the processes by the presence of a nucleus.

A: prokaryote mRNA does not undergo splicing to remove introns.

B: *translation* in prokaryotes uses N-terminal formyl methionine (F-met) as the initiator amino acid.

continued…

C: *prokaryotes* have no nucleus, the mRNA has no nuclear membrane to cross, and translation begins before its synthesis is complete.

Eukaryotic mRNA must be processed (5' G-cap, 3' poly-A tail, and splicing of exons with the removal of introns) and transported across the nuclear membrane before translation begins in the cytoplasm.

34. B is correct.

Surface-to-volume ratio limits the size of biological cells because when the volume increases, so does the surface area, but the surface area increases at a slower rate.

Eukaryotes grow beyond the apparent limitation of surface-to-volume ratios because of organelles (specialized environment) and extensive cytoplasmic matrix, utilizing microfilaments and motor proteins for cargo transport.

35. D is correct.

Prokaryotes have *peptidoglycan cell walls*, 30S and 50S ribosomes, and plasma membranes without cholesterol.

36. E is correct.

Viruses consist of genetic material with single-stranded DNA, double-stranded DNA, single-stranded RNA, or double-stranded RNA packaged within a protein coat.

A: viruses do not have membrane-bound organelles but consist entirely of nucleic acid (RNA or DNA) and, for retroviruses, reverse transcriptase (enzyme).

C: bacteria have *peptidoglycan* (i.e., N-linked glucose polymers) cell walls.

D: viruses do not have a phospholipid bilayer membrane.

37. A is correct.

Episome is an F^- plasmid that can integrate into the bacterial chromosome by homologous recombination.

38. B is correct.

Reverse transcriptase (enzyme) converts RNA of the virus into ds-DNA when infected into host cells.

The host cell treats the viral DNA as part of its genome, translating and transcribing the viral genes and the cell's genes, producing the proteins required to assemble new copies of the virus.

Newly synthesized DNA is incorporated into the host cell genome by integrase (enzyme), at which point the retroviral DNA is a *provirus*.

Retrovirus flow of genetic information:

RNA (virus) → DNA (from virus template) → RNA (host cell machinery) → polypeptide

39. A is correct.

T4 has a DNA genome transcribed and translated by the host's machinery.

Transcription and translation occur in the cytoplasm since the bacterial cell does not contain a nucleus.

Therefore, both processes occur in the same locations and are concurrent.

B: a late gene encodes lysozyme enzymes because the host is not lysed until the viruses are assembled and packaged into their protein capsids to be released for infection of other host cells.

C: the assembly must be complete before lysis.

D: bacterial cells have a *peptidoglycan cell wall*, so budding cannot occur.

The cell wall is lysed by lysozyme, and the host cell ruptures, which causes the release of the virus.

40. C is correct.

Gram-positive bacteria take up the violet stain used in the Gram staining.

Differential staining distinguishes them from the other large group of Gram-negative bacteria that cannot retain the crystal violet stain.

Gram-negative bacteria take up the counterstain (e.g., safranin, fuchsine) and appear red or pink.

Gram-positive bacteria retain the crystal violet stain due to a *thick peptidoglycan layer* superficial to the cell membrane.

In Gram-negative bacteria, this peptidoglycan layer is much thinner between cell membranes.

The cell membrane, peptidoglycan layer, and cell wall are three distinct structures.

Cell walls provide structural support, protection, and rigidity to the cell.

D: *teichoic acids* are in the cell walls of Gram-positive bacteria (e.g., *Staphylococcus, Streptococcus, Bacillus, Clostridium, Listeria*) and extend to the surface of the peptidoglycan layer.

Teichoic acids are not in Gram-negative bacteria.

41. C is correct.

A bacterium with an outer lipopolysaccharide layer is Gram-negative and protects against penicillin.

A: *fimbriae* (i.e., a proteinaceous appendage in many Gram-negative bacteria) is thinner and shorter than a flagellum and allows a bacterium to attach to solid objects.

B: *bacterial cell membranes* have a phospholipid bilayer like eukaryotes, except it lacks cholesterol.

D: *Gram-negative bacteria* have a thinner peptidoglycan cell wall that does not retain Gram stain.

42. E is correct.

DNase is an enzyme that degrades DNA by hydrolysis of the DNA molecule.

Bacterial cells treated with DNase die because bacteria have DNA genomes, while RNA viruses are unaffected by DNase and continue to synthesize proteins following treatment with DNase.

A: having multiple copies of a gene is not enough to prevent DNase from degrading the DNA.

C: *viral genomes* typically contain multiple reading frames to use their limited nucleic acid efficiently, but multiple reading frames do not prevent DNA hydrolysis.

D: viral protein coat is not able to denature DNase.

Denaturation breaks weak (e.g., hydrogen, dipole, and hydrophobic) bonds by heat or chemical treatment.

43. C is correct.

F^- recipient remains F^- in mating between Hfr and F^- cells.

44. A is correct.

Translation in prokaryotes uses the *initiation* amino acid of N-terminal formyl methionine (F-met).

Specific cells in the human immune system can identify f-MET and release local toxins to inhibit bacterial (i.e., prokaryote) infections.

45. C is correct.

Dextran is a large polysaccharide and does *not pass* through the cell membrane.

Osmotic pressure of the solution increases, H_2O moves out of the cell to reduce the osmotic pressure difference, and the cell undergoes crenation (i.e., shrinking).

46. A is correct.

Gram staining differentiates bacterial species into two large groups: Gram-positive and Gram-negative.

Gram staining differentiates the chemical and physical properties of *peptidoglycan cell walls.*

Gram-positive bacteria have a thicker peptidoglycan cell wall which gives Gram-positive bacteria a purple-blue color from staining, while Gram-negative result in a pink-red color.

Gram staining is usually the first assay in identifying a bacterial organism.

47. A is correct.

DNA polymerase replicates DNA strands by synthesizing new strands from a template strand.

DNA polymerase replicates the DNA F factor in F^+ cells before conjugation.

B: *reverse transcriptase* is a retroviral enzyme synthesizing DNA from an RNA (virus) template.

C: *DNA ligase* catalyzes the formation of phosphodiester bonds that link adjacent DNA bases.

D: *integrase* is a retroviral enzyme that integrates provirus DNA into host genomes.

48. B is correct.

Conjugation occurs between bacterial cells of different mating types.

"Maleness" in bacteria is determined by the presence of a small extra piece of DNA that can replicate independently of the larger chromosome.

Male bacteria having this *sex factor* (the *F factor*) are denoted F^+ if the sex factor exists as extrachromosomal.

F^+ bacteria can conjugate only with F^- bacteria, the "female" that do not possess the F factor.

Genes on the F factor determine the formation of *sex pili* (hair-like projections) on the surface of the F^+ bacterium, which forms cytoplasmic bridges to transfer genetic material. The pili aid the F^+ cell in adhering to the F^- cell during conjugation.

During conjugation (bacterial mating) of an F^+ cell with an F^- cell, and before the transfer, the F factor replicates, and the F factor is the DNA likely to be transferred to the female. F^- becomes an F^+ by receiving one copy of the F factor, while the original F^+ retains a copy.

If this were the only genetic exchange in conjugation, all bacteria would become F^+ and cease conjugation. In F^+ bacterial cultures, a few bacteria with the F factor incorporated into their chromosome can be isolated and are *Hfr* bacteria, which may conjugate with F^- cells.

They do not transfer their F factor during conjugation but often transfer linear portions of their chromosomes; the transfer is interrupted by the spontaneous breakage of the DNA molecule at random sites, usually, before the F factor crosses to the F^- cell.

This process is *unidirectional*, and no genetic material from the F^- cell is transferred to the *Hfr* cell.

49. E is correct.

Retrotransposons (or transposons via RNA intermediates) are a subclass of transposons. They are endogenous genetic elements that amplify in a genome as ubiquitous components of DNA in many eukaryotic organisms.

Around 42% of the human genome is retrotransposons.

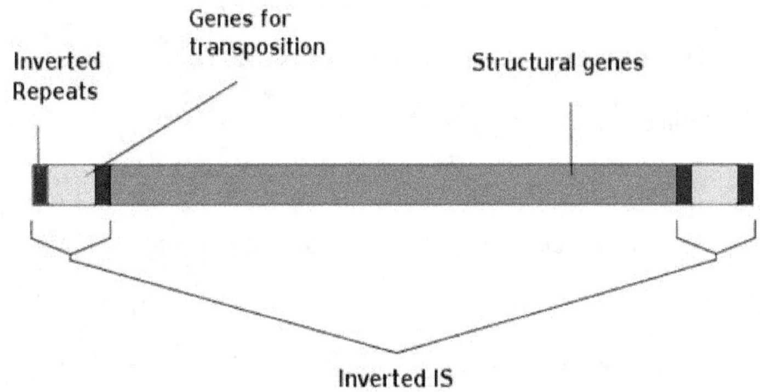

Bacterial DNA transposon with flanking inverted sequences encompassing structural genes

Retrotransposons are copied into RNA and back into DNA which may integrate into the genome.

The second step of forming DNA may be catalyzed by a reverse transcriptase encoded by the retrotransposon.

50. B is correct.

Plasmids are small circular double-stranded DNA in bacteria carrying extrachromosomal genetic information.

Plasmids (with an origin of replication) are replicated by bacterial proteins and inherited by progeny.

Proteins encoded by genes on plasmids often provide resistance to antibiotics by degrading the antibiotic.

Through recombinant DNA technology, plasmids are engineered to carry genes not typically in bacteria.

Plasmids are introduced into cells by transforming DNA across the cell wall and plasma membrane without killing bacterial cells.

After transformation, exposure to the specific antibiotic allows the selection of bacteria that received plasmid.

Bacteriophages, such as bacteriophages λ are viruses that infect bacterial cells.

Bacteriophage λ are engineered to carry novel genes, not in bacteria or bacteriophages.

I: *plasmids* are extra-chromosomal circular DNA molecules and not organelles because organelles are membrane-bound cellular components present only in eukaryotes.

II: like the bacterial genome, plasmids are in the cytoplasm.

Without a nucleus, prokaryotic ribosomes translate plasmid mRNA into proteins while transcribed into mRNA from DNA.

III: plasmids rely on bacterial machinery for metabolic processes (e.g., replication, transcription, translation).

51. B is correct.

Fungi are members of a large group of eukaryotic organisms that includes microorganisms such as yeasts, molds, and mushrooms.

These organisms are classified as a Fungi kingdom (separate from plants, animals, protists, and bacteria).

One major difference is that fungal cells have cell walls with *chitin.*

The cell walls of plants contain *cellulose*), some protists contain *cellulose,* and bacteria contain *peptidoglycan.*

A: *protoplasts* are plant, bacterial or fungal cells that lose their cell wall by mechanical or enzymatic means.

C: L forms are strains of bacteria that lack a cell wall.

D: viruses do not have cell walls.

A virus particle (i.e., virion) has nucleic acid enclosed in a capsid as a protective protein coat.

E: *mycoplasma* is a genus of bacteria that lack a cell wall (they have cell membranes only).

Mycoplasma are unaffected by many common antibiotics, such as penicillin or other beta-lactam antibiotics, targeting peptidoglycan cell wall synthesis without a cell wall.

52. E is correct.

Prokaryotes do *not* have membrane-bound organelles (e.g., nuclei, mitochondria, lysosomes).

Prokaryotes have *peptidoglycan* cell walls.

Fungi have cell walls with *chitin.*

53. C is correct.

Bacteriophages are viruses that infect bacteria and typically consist of a protein coat (i.e., head) and a core containing nucleic acid.

Like viruses, bacteriophages contain host-specific protein tail fibers for attachment.

Upon infection, bacteriophages can enter one of two life cycles (image below).

In the *lytic cycle*, the viral nucleic acid enters the bacterial cell.

Lytic cycle begins using host machinery to produce new virions (i.e., virus particles), lysing the host cell and infecting other cells.

Lysogenic cycle is when viral DNA integrates into the bacterial chromosome, replicating with it and being passed to daughter cells (integrated into their genome) in this inactive form.

Lytic cycle destroys the infected cell and its membrane.

A critical difference between the lytic and lysogenic phage cycles is that in the lytic phage, the viral DNA exists as a separate molecule within the bacterial cell and replicates separately from the host bacterial DNA.

The location of viral DNA in the lysogenic phage cycle is within the host DNA (i.e., integrated); the virus (phage) replicates using the host DNA machinery.

Phage is a free-floating separate molecule to the host DNA in the lytic phage cycle.

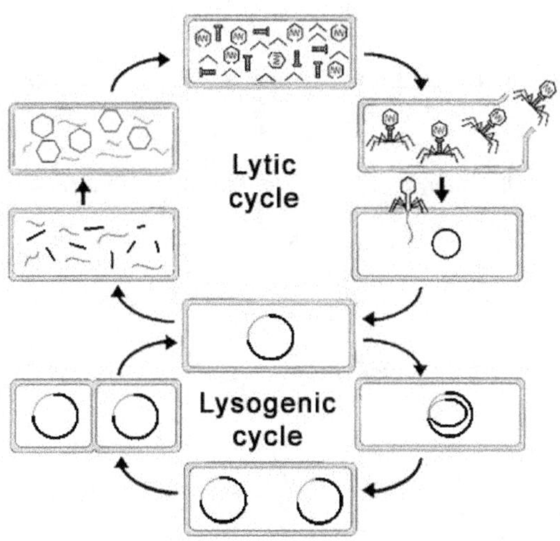

Lytic phase lysis the cell and releases virions while lysogenic cycles involve the integration of the viral genome into the host cell

continued…

When an integrated virus (i.e., prophage) becomes activated, it exits the lysogenic and enters the lytic cycle.

Activation triggers for the virus to enter the lytic cycle include *cellular or metabolic stresses* such as UV light, depleted nutrients, or cellular abnormalities.

Retroviruses are RNA-containing viruses replicating by reverse transcriptase (enzyme) through a DNA intermediate via a viral-coded *RNA-dependent DNA polymerase*.

For example, human immunodeficiency virus (HIV), the causative of AIDS, is a retrovirus.

Retroviral life cycle consists of four main events:

1) virus binds its host and injects its RNA and a few viral enzymes;

2) RNA is converted to DNA by reverse transcriptase;

3) DNA integrates into the host cell's genome; and

4) viral genes are expressed, and virions are assembled and released from the cell by budding.

Bacteriophage and retrovirus integrate into the host genome (i.e., lysogenic) or lyse cells (i.e., lytic cycle).

A: *bacteriophage*s have host-specific protein tail fibers and infect bacteria.

B: bacteria (the target of bacteriophage) have no immune system.

Bacteria have a defense against infection by nucleotide-specific *restriction enzymes* used for molecular biology and biotechnology.

D: *retroviruses* contain an RNA genome (ss-RNA or ds-RNA) and reverse transcriptase, not in bacteriophages.

54. E is correct.

Integration of phage DNA into bacterial chromosomes uses *site-specific recombinase* encoded in the phage.

55. A is correct.

Sequences recognized by most restriction enzymes are *inverted repeats* (i.e., palindromes) that read the same if inverted (i.e., rotated by 180°).

Sticky ends for DNA fragments are the same in either orientation, so ligation occurs from either orientation.

B: DNA strands do serve as primers for DNA polymerase

C: DNA *ligase* creates a phosphodiester bond that covalently links sticky ends of DNA.

D: *plasmid DNA*, like bacterial DNA, is double-stranded.

56. D is correct.

Virulence describes disease severity or a pathogen's infectivity.

The virulence factors of bacteria are typically proteins or other molecules synthesized by enzymes.

Virulent viruses use the *lytic* lifecycle.

Proteins are coded by genes in chromosomal DNA, bacteriophage DNA or plasmids.

57. C is correct.

Promoter is a region on DNA recognized and bound by RNA polymerase as the initiation site for transcription.

Inducer is the molecule that inactivates the repressor.

When the inducer binds the *repressor*, the bound repressor dissociates from DNA, allowing polymerase to bind and activate the operon.

Repressor binds the *operator* (i.e., a segment of DNA within the operon).

58. A is correct.

Genetic marker is a gene or DNA sequence with a known location on a chromosome used to identify individuals.

It can be a variation that results from mutation or alteration in the genomic loci.

Genetic marker may be a short DNA sequence, such as a sequence surrounding a single base-pair change (single nucleotide polymorphism – SNP) or a long one (e.g., minisatellites).

59. E is correct.

Transformation is "horizontal gene transfer" when foreign genes transfer among bacteria.

In transformation, foreign DNA is taken *from the medium* and incorporated directly through the cell membrane.

Transformation depends on the recipient bacterium, which must be in a state of natural or artificial competence to take up exogenous genetic material.

Two other mechanisms for *horizontal gene transfer* are *conjugation* (i.e., genetic material between bacteria in direct contact) and *transduction* (i.e., injection of exogenous DNA by a bacteriophage into the host bacterium).

60. C is correct.

Yeasts are *eukaryotic model systems* for molecular biology studies as they exhibit fast growth and have dispersed cells. They have a well-defined genetic and versatile DNA transformation system that can be utilized for protein production.

Transformation occurs when exogenous DNA is introduced into a cell, resulting in an inheritable change or genetic modification, as first reported in *Streptococcus pneumoniae* by Griffith in 1928.

DNA transformation was demonstrated by Avery *et al.* in 1944.

Budding yeast *Saccharomyces cerevisiae* was first successfully transformed in 1978.

Plasmids are prokaryotic (i.e., bacteria) and do *not* undergo post-translational modification (i.e., removal of introns and ligating exons). Plasmids are smaller strands of nucleotides than eukaryotic chromosomes.

Yeast artificial chromosome (YAC) is a human-engineered DNA molecule used to clone deoxynucleic acid sequences in yeast cells.

YACs are engineered with DNA segments of one million base pairs for mapping and sequencing genomes.

Notes for active learning

Notes for active learning

Notes for active learning

Section III: DNA, Genes & Heredity

Mechanisms of Reproduction – Detailed Explanations

1. B is correct.

Gametes form via *meiosis* and are double-stranded haploids.

A single chromosome consists of two hydrogen-bonded complementary antiparallel DNA strands.

2. C is correct.

Let C designate wild-type and c designate the colorblind allele.

Mother is Cc, and the father is CY (a normal allele with a single copy of the X gene).

From mating, the mother's gamete (as a carrier due to her dad) is C or c, with a 50% probability of the gamete inheriting the C or c allele.

Assuming a boy (i.e., the father transmits Y and not X), the father's allele of Y (boy) is 100%, and the probability of the mother passing a c (colorblind) is 50%.

Probability of a son being colorblind is 50% or ½.

If the question had asked, "what is the probability that they will have a colorblind child?" the analysis changes to determine the probability for all children (not just boys in the original question).

Gametes produced by the mother are C and c with a 50% probability each.

Affected child is a boy.

What is the probability that the father passes the X or Y gene to the offspring?

Probability is 50%.

Individual event probabilities are multiplied to determine the overall probability:

$\frac{1}{2} \times \frac{1}{2} = \frac{1}{4}$

3. E is correct.

Primary oocytes are arrested in meiotic prophase I from birth until ovulation within the ovaries.

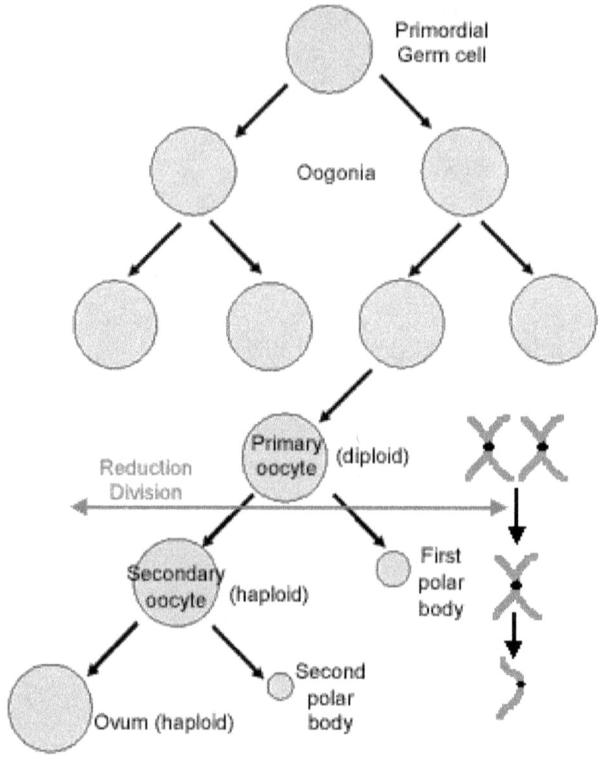

Diploid primordial germ cell undergoes two rounds of meiosis to form haploid ovum and polar bodies

4. C is correct.

Seminiferous tubules are in the testes and are the site of sperm production. *Spermatozoa* are the mature male gametes in many sexually reproducing organisms. *Spermatogenesis* is how spermatozoa are produced from male primordial germ cells by mitosis and meiosis.

Initial cells in this pathway are spermatogonia, which yield 1° spermatocytes by mitosis.

1° spermatocyte divides via meiosis into two 2° spermatocytes.

Meiosis converts 1° spermatocyte (2N) into four spermatids (1N).

2° spermatocytes complete meiosis by dividing into 2 spermatids as mature spermatozoa (i.e., sperm cells).

1° spermatocyte gives rise to two 2° spermatocytes that, by further meiosis, produce four spermatozoa.

Seminiferous tubules are in the testes and the location of meiosis and subsequent creation of gametes (e.g., spermatozoa for males or ova for females).

Epithelium of the tubule consists of *Sertoli cells* whose primary function is to nourish the developing sperm cells through the stages of spermatogenesis.

Sertoli cells function as phagocytes, consuming the residual cytoplasm during spermatogenesis.

Spermatogenic cells differentiate between the Sertoli cells, which differentiate through meiosis into sperm cells.

5. C is correct.

A human cell after the first meiotic division is 1N with 2 chromatids.

6. D is correct.

Genetic recombination is when two DNA strand molecules exchange genetic information (i.e., a base composition within the nucleotides), resulting in new combinations of alleles (i.e., alternative forms of genes).

In eukaryotes, the natural process of genetic recombination during meiosis (i.e., the formation of gametes – eggs and sperm) results in genetic information passed to progeny.

Genetic recombination in eukaryotes involves pairing homologous chromosomes (i.e., a set of maternal and paternal chromosomes), which may involve nucleotide exchange between the chromosomes.

Information exchange may occur without physical exchange when a section of genetic material is duplicated without a change in the donating chromosome or by breaking and rejoining the DNA strands (i.e., forming new molecules of DNA).

Mitosis may involve recombination, where two sister chromosomes form after DNA replication.

New combinations of alleles are often not produced because the sister chromosomes are usually identical.

In meiosis and mitosis, recombination occurs between similar DNA molecules (homologous chromosomes or sister chromatids, respectively).

In meiosis, non-sister (i.e., same parent) homologous chromosomes pair with each other, and recombination often occurs between non-sister homologs.

For somatic cells (i.e., undergo mitosis) and gametes (i.e., undergo meiosis), recombination between homologous chromosomes or sister chromatids is a common DNA repair mechanism.

7. C is correct.

Primary oocytes (2N cells) are arrested in prophase I of meiosis I until puberty. There is additional development into secondary oocytes that occurs prior to ovulation.

After ovulation, the oocyte is arrested in metaphase of *meiosis II until fertilization.*

8. B is correct.

Pseudoautosomal regions are named because any genes within them are inherited, like autosomal genes.

Pseudoautosomal regions allow males to pair and segregate X and Y chromosomes during meiosis.

Males have two copies of genes: one in the pseudoautosomal region of their Y chromosome and the other in their X chromosome's corresponding portion.

Typical females possess two copies of pseudoautosomal genes, as each X chromosome contains pseudoautosomal regions.

Crossing over (during prophase I) between the X and Y chromosomes is usually restricted to the pseudoautosomal regions.

Pseudoautosomal genes exhibit an *autosomal*, rather than sex-linked, inheritance pattern.

Females can inherit an allele initially on the Y chromosome of their father, and males can inherit an allele initially on the X chromosome of their father.

9. E is correct.

Primary spermatocyte completes the synthesis (S) phase of interphase, not the first meiotic division, and remains diploid (2N) with 46 chromosomes (i.e., 23 pairs).

10. B is correct.

Chromosomes (not chromatids) *segregate* during mitosis to produce identical somatic (body) 2N (diploid) cells from the parental cell. Meiosis is the process for germline cells (i.e., egg, sperm).

Mutations are inheritable changes in the cell's genetic (DNA) material; recombination occurs during prophase I (at the chiasma of the tetrad) of meiosis, and homologous chromosomes segregate during meiosis I.

During meiosis I, homologous chromosomes separate. Subsequently, during meiosis II, the sister chromatids separate to produce four 1N (haploid) products, each with half the number of chromosomes as the original cell.

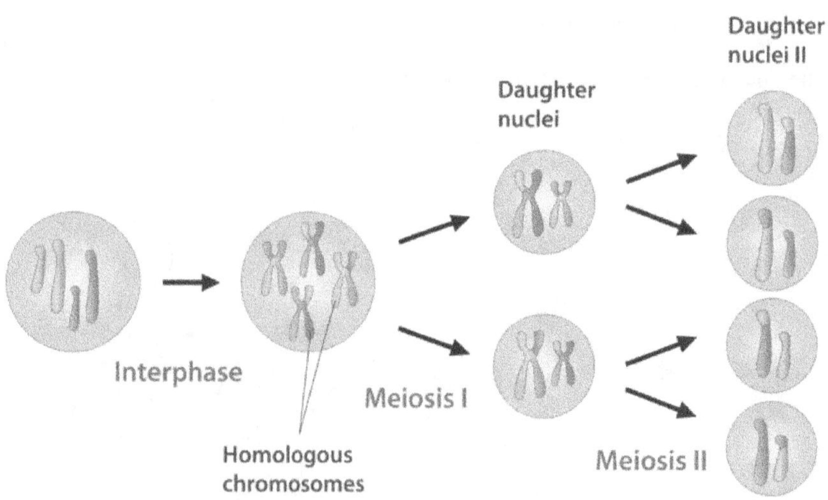

Chromosomes replicate in S phase of mitosis to produce homologous chromosomes. Meiosis I daughter cells are diploid (2N), while the gametes formed from meiosis II (reductive phase) are haploid (1N).

11. E is correct.

In females, secondary oocytes are haploid (i.e., single copies of 23 chromosomes, each with pair of chromatids).

Secondary oocytes do not complete meiosis II (i.e., haploid with a single chromatid) until fertilized by sperm.

Each month during puberty, one primary oocyte (i.e., diploid, each with a pair of chromatids) completes meiosis I to produce a secondary oocyte (1N) and a polar body (1N).

The 1N secondary oocyte (i.e., 23 chromosomes each with a pair of chromatids) is expelled as an ovum from the follicle during ovulation, but meiosis II does not occur until fertilization.

Menarche is marked by the onset of menstruation and signals the possibility of fertility.

Menstruation is the sloughing of the *endometrial lining* of the uterus during the monthly hormonal cycle.

Menopause is when menstruation ceases.

12. D is correct.

Klinefelter syndrome describes the symptoms from additional X genetic material in males.

Turner syndrome is the condition in females with a single X (monosomy X) chromosome.

XYY syndrome is a genetic condition in which a human male has an extra male (Y) chromosome, giving 47 chromosomes instead of 46. XYY is not inherited but usually occurs during the formation of sperm cells.

Nondisjunction error during anaphase II (meiosis II) results in sperms with an extra Y chromosome.

If an atypical sperm contributes to genetic makeup, the child has an extra Y chromosome in each somatic cell.

Triple X syndrome is not inherited but usually occurs during the formation of gametes (e.g., ovum and sperm) because nondisjunction in cell division results in reproductive cells with additional chromosomes.

During cell division, errors (non-disjunction) can result in gametes with additional chromosomes.

An egg or sperm may gain an extra X chromosome due to non-disjunction, and if one gamete contributes to the zygote, the child will have an extra X chromosome in each cell.

13. C is correct.

In males, diploid spermatogonia (2N) cells undergo meiosis I to produce 1° diploid spermatocyte (2N), which undergoes meiosis II to yield 2° haploid spermatocytes (1N).

2° spermatocytes undergo meiosis II to produce four spermatids (1N) that mature into spermatozoa (i.e., sperm).

In females, 1° oocyte is (2N), whereby 2° oocyte undergoes the second meiotic division to produce two (1N) cells – a mature oocyte (i.e., ovum – female gamete) and another polar body.

Primary oocyte is diploid (2N).

During fertilization, a (1N) ovum and a (1N) sperm fuse to produce a (2N) zygote.

Spermatogonium is a diploid (2N) cell.

continued…

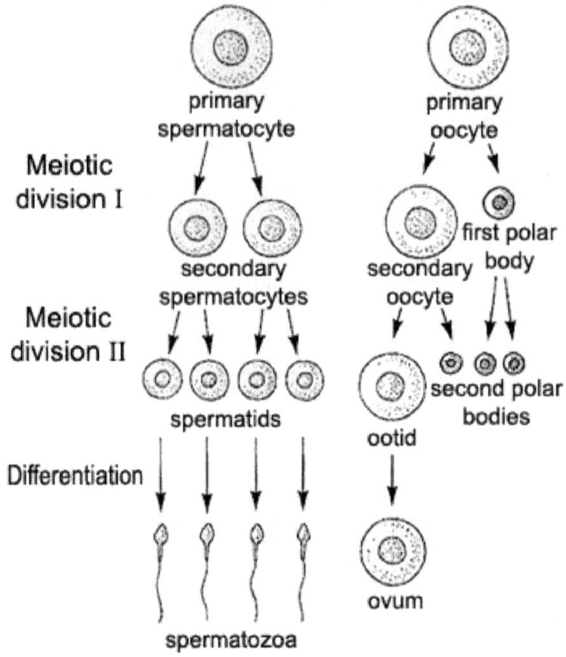

Spermatogenesis Oogenesis

Diploid primary oocytes and oocytes undergo two rounds of meiosis forming haploid spermatozoa and ovum

14. D is correct.

Probability that a child will be male (or female) is ½.

Each event is *independent*; the first (or second or third) child does *not* affect future events.

Probability for 4 children is:

$$\frac{1}{2} \times \frac{1}{2} \times \frac{1}{2} \times \frac{1}{2} = 1/16$$

Same probability if the first and third child are female, with the second and fourth as male (or combination).

15. D is correct.

Unequal division of cytoplasm occurs during the meiotic process of oogenesis (i.e., production of an egg cell). Meiotic divisions occur in two stages.

First stage produces precursor (2N) cells when a daughter cell (precursor egg) receives most of the cytoplasm; the other is devoid of sufficient cytoplasm (but genetically identical) and a nonfunctioning polar body.

Second division of oogenesis is when the large (2N) daughter cell divides again, and again one (1N) daughter cell receives the most cytoplasm (i.e., putative *ovum*).

The other daughter eggs become small nonfunctional (1N) *polar bodies*.

Polar bodies (from meiosis I) may divide (via meiosis II) to form two nonfunctional (1N) polar bodies.

The result is a potential of four (1N) cells.

continued…

However, only one—the one with a greater amount of cytoplasm during each meiotic division—becomes a functional egg (i.e., ovum) cell along with (up to) three (1N) polar bodies.

Bacterial cells divide for reproduction, and cytoplasmic division is equal.

Mitosis of kidney cells distributes cytoplasm equally.

Spermatogenesis is when one (2N) precursor cell forms four functional (1N) sperm cells because both meiosis divisions are equal and contain an equal amount of cytoplasm.

16. B is correct.

Dyneins are motor proteins (or *molecular motors*) using ATP to perform movement as *retrograde transport.*

Dynein transports cellular contents by *walking* along cytoskeletal microtubules towards the *minus-end* of the microtubules, usually oriented towards the cell center (i.e., *minus-end directed motors.*)

Kinesins are motor proteins that move toward the *plus-end* of the microtubules (i.e., *plus-end directed motors*).

Ovulation is the phase of the female menstrual cycle. A partially mature ovum that has yet to complete meiosis II is released from the ovarian follicles into the Fallopian tube (i.e., oviduct).

After ovulation, the egg can be fertilized by sperm during the *luteal phase.*

Ovulation is determined by circulating hormone levels and is not affected by a defect in dynein motor proteins.

Lungs require cilia to remove bacteria and other particulates.

Kartagener's syndrome males would be *infertile due to sperm immobility.*

Ova would not typically enter the Fallopian tubes (i.e., oviduct) because of the lack of cilia, which would cause an increased risk of ectopic pregnancy.

E: *epithelium lining* the tube of the middle ear is ciliated.

Beat is directed from the cavity to the pharynx, which clears mucus and pathogens.

Therefore, interference with the ciliary function would increase middle ear infections.

17. E is correct.

DNA replication occurs during the synthesis (S) phase of interphase (i.e., G1, S, G2) to form *sister chromatids* joined by a centromere.

Replication occurs during the cell cycle *S* phase (i.e., interphase).

Transcription (in the nucleus) and *translation* (in the cytoplasm) occur during the S phase.

18. C is correct.

Progesterone is a steroid hormone involved in the female menstrual cycle, pregnancy (supports *gestation*), and embryogenesis.

Progesterone levels are relatively low in women during the pre-ovulatory phase of the menstrual cycle, rise after ovulation, and are elevated during the luteal phase.

During pregnancy, human chorionic gonadotropin (HCG) is released, maintaining the corpus luteum and allowing it to maintain progesterone levels.

At 12 weeks, the placenta begins to produce progesterone in place of the corpus luteum – this process is the *luteal-placental shift*.

After the luteal-placental shift, progesterone levels start to rise further.

After delivery of the placenta and during lactation, progesterone levels are very low.

Progesterone levels are low in children and postmenopausal women.

Adult males have levels like those in women during the follicular phase of the menstrual cycle.

19. A is correct.

During meiosis, the gamete (e.g., ovum and sperm) reduces its genetic component from diploid (2N) to haploid (1N) with half the typical chromosome number for somatic (i.e., body cells).

When a haploid egg and sperm unite, they form a diploid zygote.

Ova contains an X chromosome, while sperm contains an X or a Y chromosome.

Gametes form in meiosis (i.e., two reduction divisions) without intervening chromosome replication.

During prophase I of meiosis I, *tetrads* form, and *sister chromatids* (i.e., a chromosome replicated in a prior S phase) undergo homologous recombination as *crossing over*.

Crossing over increases genetic variance within the progeny and is a driving force in the evolution of a species.

20. E is correct.

Meiosis is cell division in sexually reproducing eukaryotes (animals, plants, and fungi), whereby the chromosome number is reduced by half, resulting in four genetically distinct haploid daughters.

DNA replication is followed by mitosis and *two rounds of cell division* in meiosis to produce four (1N) cells.

Two rounds of meiotic divisions are Meiosis I and Meiosis II.

Meiosis I is a *reductive division*, as the cells are reduced from diploid (2N) to haploid (1N).

Meiosis II (only G phase separates I and II) is an *equational division*, as the cells begin and end as haploids.

21. D is correct.

Crossing over occurs during prophase I of meiosis. During prophase I, the *chromatin condenses* into chromosomes, the centrioles migrate to the poles, the spindle fibers begin to form, and the nucleoli and nuclear membrane disappear.

Homologous chromosomes physically pair and intertwine in the process of *synapsis.*

Prophase I: *chromatids* (i.e., a strand of the replicated chromosome) of *homologous chromosomes* break and exchange equivalent pieces of DNA via *crossing over.*

Metaphase I: *homologous pairs align* at the equatorial plane, and each pair attaches to a separate spindle fiber by its *kinetochore* (i.e., protein collar around the centromere of the chromosome).

Anaphase I: *homologous chromosomes separate* and are pulled by the spindle fibers to opposite cell poles.

Disjunction (i.e., separation of homologous chromosomes) is essential for segregation, as described by Mendel.

Telophase I: *nuclear membrane forms* around each new nucleus, and chromosomes consist of sister chromatids joined at the centromere.

The cell divides into two daughter cells, each receiving a haploid nucleus (1N) of chromosomes.

Interkinesis is a brief period between the two reduction cell divisions of meiosis I and II and during which the chromosomes partially uncoil.

22. A is correct.

In prophase I, *tetrads form*, genetic recombination occurs, and the spindle apparatus forms.

Chromosomes migrate to the poles of the cell during anaphase from the *splitting of the centromere.*

23. E is correct.

Spermatogenesis and *oogenesis* are *gametogenesis.*

Haploid (1N) gametes (i.e., ova and sperm) are produced by diploid (2N) cell reductive divisions (i.e., meiosis).

Spermatogenesis occurs in the gonads, whereby the cytoplasm equally divides during meiosis with the production of four viable sperm 1N cells.

Oogenesis occurs in the gonads, whereby the cytoplasm divides unequally. One 1N ovum (e.g., egg) receives the bulk of the cytoplasm and (up to) three additional 1N polar bodies.

Polar bodies contain a 1N genome (like sperm and egg) but lack sufficient cytoplasm for a viable gamete.

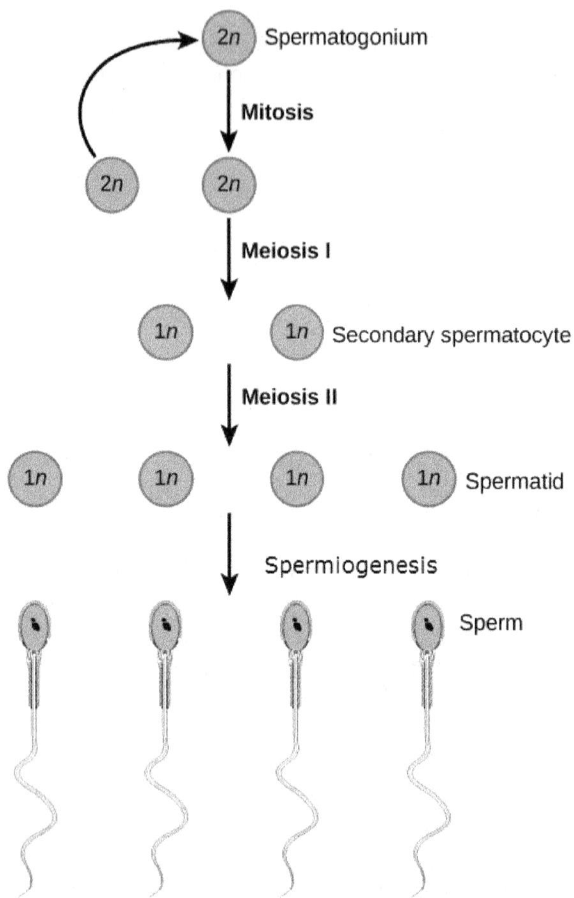

Diploid primary oocytes and oocytes undergo two rounds of meiosis forming haploid spermatozoa and ovum

24. B is correct.

Barr body is an *inactive X chromosome* in a female somatic cell rendered inactive by *lyonization* for species in which sex is determined by the Y chromosomes (i.e., humans and other species).

Lyon hypothesis states that cells with multiple X chromosomes inactivate all but one X randomly during mammalian embryogenesis.

In humans with more than one X chromosome, the number of Barr bodies visible at interphase is one less than the total number of X chromosomes.

For example, a man with Klinefelter syndrome of 47, XXY karyotype has a single Barr body.

A woman with a 47, XXX karyotype has two Barr bodies.

25. A is correct.

Polar bodies form in females as 1N, nonfunctional cells during meiosis.

Meiosis is a two-stage process whereby a 2N cell undergoes a reduction division to form two 1N cells.

Each 1N cell forms two 1N cells for four 1N cells from a parental 2N germ cell in the *second meiotic* division.

For sperm cells, four functional, unique 1N gamete sperm cells form.

For egg cells, the first meiotic division involves an *unequal division of cytoplasm* that results in one large cell and one small cell (i.e., the first polar body).

The large cell divides (second meiotic division) to generate *one functional egg* (1N ovum containing most of the cytoplasm) and another polar body.

First polar body divides again to form two other polar bodies. One large ovum (i.e., functional egg cell) and three polar bodies form during a female meiotic division.

Polar bodies form in mitosis, which is how somatic (i.e., body cells) cell divisions occur; equal cells do not form gametes (i.e., germline cells).

26. E is correct.

Turner syndrome (i.e., 45, X) describes several conditions in females, of which monosomy X (i.e., the absence of the entire sex (X) chromosome, the Barr body) is the most common.

Turner syndrome is a chromosomal abnormality when all or part of a sex chromosome is absent or abnormal.

27. D is correct.

Dolly, the sheep, was cloned in 1996 by fusing the *nucleus* from a mammary gland cell of a Finn Dorset ewe into an *enucleated egg cell* from a Scottish Blackface ewe.

During gestation, Dolly was carried to term in the uterus of another Scottish Blackface ewe.

She was a genetic copy of the Finn Dorset ewe from the somatic mammary gland cell.

Dolly's creation showed that somatic (i.e., body cell) cell DNA in a differentiated (*vs. pluripotent* or *totipotent*) could be induced through nuclear transfer (i.e., transplantation) to expand developmental fate (e.g., like a germ cell) in progression from a zygote.

Notes for active learning

Notes for active learning

Notes for active learning

Development – Detailed Explanations

1. A is correct.

Respiratory exchanges during fetal life occur through the *placenta.*

Placenta connects the developing fetus to the uterine wall and allows nutrient uptake, waste elimination, and gas exchange via the mother's blood supply.

Umbilical cord connects the developing embryo (or fetus) and the placenta. During prenatal development, the umbilical cord is physiologically and genetically part of the fetus and contains two arteries (the umbilical arteries) and one vein (the umbilical vein).

Umbilical veins supply the fetus with oxygenated, nutrient-rich blood from the placenta. Conversely, the fetal heart pumps deoxygenated, nutrient-depleted blood through the umbilical arteries back to the placenta.

Fetal circulatory systems change at birth as the newborn uses its lungs. After the infant's first breath, the newborn's cardiovascular system constricts the ductus arteriosus (i.e., connects the pulmonary artery to the aorta) and converts it to the *ligamentum arteriosum.* Resistance in the pulmonary blood vessels decreases, increasing blood flow to the lungs.

At birth, umbilical blood flow ceases, and blood pressure in the inferior vena cava decreases, causing a decrease in pressure in the *right atrium. Left atrial pressure* increases due to increased blood flow from the lungs. Increased left atrial pressure, coupled with decreased right atrial pressure, causes *closure* of the *foramen ovale.*

Ductus venosus, which shunts blood from the left umbilical vein directly to the *inferior vena cava* to allow oxygenated blood from the placenta to bypass the liver, completely closes within three months after birth.

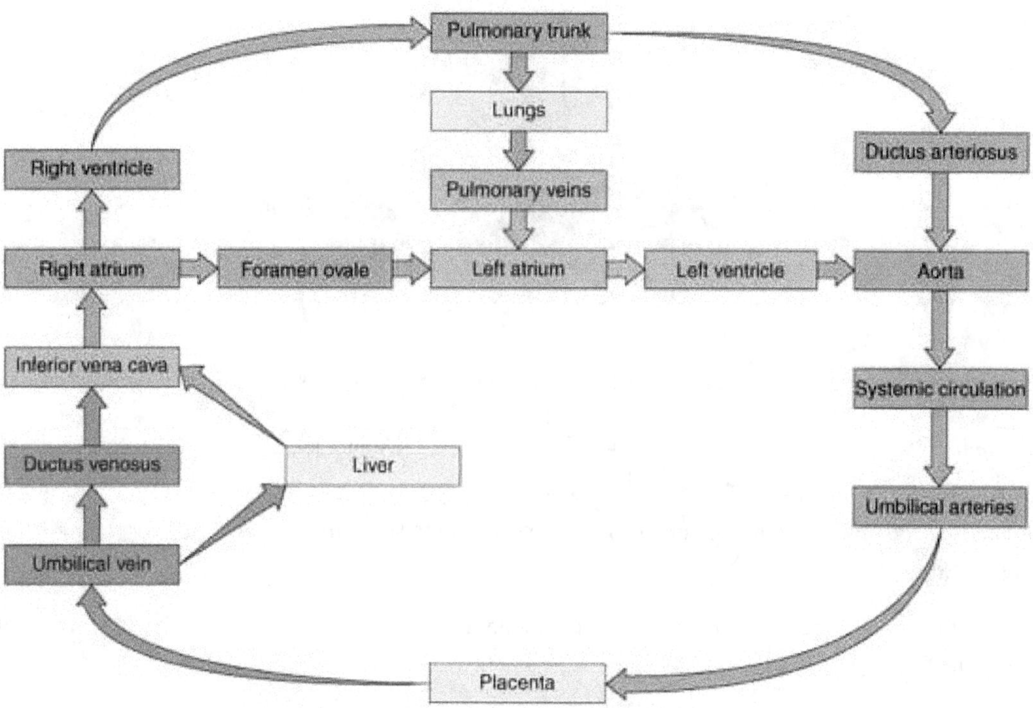

Adult and fetal circulatory systems with ductus arteriosus and ductus venosus provide shunt pathways

2. D is correct.

Indeterminate cleavage cells maintain the ability to develop into a complete organism.

Zygote implantation into the uterus causes cell migration transforming the blastula from a single-cell layer into a three-layered gastrula (i.e., ectoderm, mesoderm, and endoderm).

Blastulation begins when the morula develops the blastocoel as a fluid-filled cavity that (by the fourth day) becomes the blastula as a hollow sphere of cells.

Determinate cleavage results in cells whose differentiation potential is determined early in development.

3. B is correct.

Mutations in *Drosophila*, transforming one body segment into a different one, are in homeotic genes.

Homeotic genes encode the related homeodomain proteins involved in developmental patterns and sequences.

Homeotic genes influence the development of specific structures in plants and animals, such as the Hox and ParaHox genes, which are essential for segmentation.

Homeotic genes determine where, when, and how body segments develop. For example, alterations in homeotic genes in laboratory flies cause changes in patterns of body parts, sometimes producing dramatic effects, such as legs growing in place of antennae or an extra set of wings.

Hox genes are *homeotic transcription factors* are crucial for controlling the body plan along the *anterior-posterior axis* (i.e., craniocaudal axis) and specify the segment identity of tissues within the embryo.

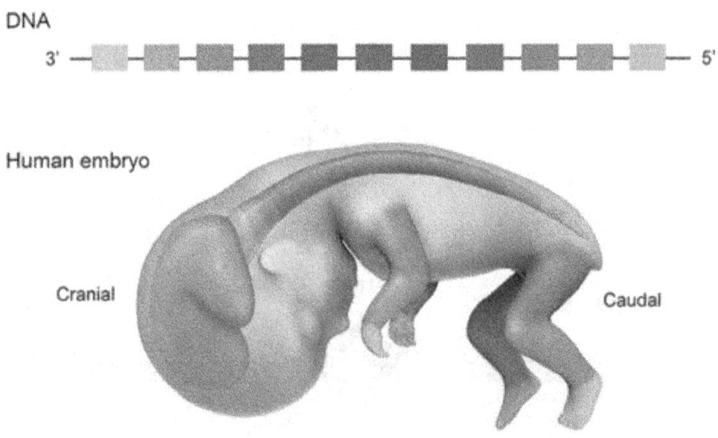

HOX genes encode homeodomain proteins as developmental regulators of the anterior-posterior axis

Homeodomain is a protein structural domain that binds DNA or RNA and is a transcription factor (i.e., facilitates polymerase binding and activation).

Homeodomain proteins fold with a 60-amino acid helix-turn-helix structure in which the short loop regions connect three alpha-helices.

N-terminus of the two helices is *antiparallel,* and the longer C-terminal helix is roughly perpendicular to the axes established by the first two.

Third helix functions as a *transcription factor* and interacts directly with DNA.

4. E is correct.

Ectoderm gives rise to hair, nails, skin, brain, and nervous system.

Ectoderm forms "outer linings," including the epidermis (i.e., outermost skin layer) and hair.

Ectoderm is the precursor to *mammary glands* and *central* (CNS) and *peripheral nervous systems* (PNS).

Connective tissue is derived from *mesoderm.*

Rib cartilage is derived from *mesoderm.*

Epithelium of the digestive system is derived from *endoderm.*

5. D is correct.

Chorion is a membrane between the developing fetus and mother in humans and most mammals.

Chorion consists of two layers:

outer layer formed by the *trophoblast*

inner layer formed by the *somatic mesoderm* (in contact with the amnion).

6. A is correct.

The embryo does *not* support the maintenance of the *corpus luteum* when an embryo lacks the synthesis of human chorionic gonadotropin (hCG).

Human chorionic gonadotropin interacts with a receptor on the ovary, producing progesterone secretion.

Progesterone creates a thick lining of blood vessels and capillaries in the uterus to sustain the fetus.

Thus, the corpus luteum is supported during pregnancy.

7. C is correct.

Mesoderm develops into the circulatory, musculoskeletal, and excretory systems, outer coverings of internal organs, gonads, and various types of muscle tissue.

Ectoderm develops into the brain and nervous system, hair and nails, lens of the eye, inner ear, sweat glands, the lining of the nose and mouth, and skin epidermis.

Endoderm develops into the epithelial lining of the digestive tract, respiratory tracts, lining of the liver, bladder, pancreas, thyroid, and alveoli of the lungs.

Epidermis is not an embryonic germ layer but the layer of the skin covering the dermis.

8. B is correct.

Gastrulation is a phase early in animal embryonic development, during which the single-layered blastula is reorganized into a trilaminar (*three-layered*) structure of the *gastrula*.

Three germ layers are:

ectoderm, *mesoderm,* and *endoderm.*

Gastrulation occurs after cleavage, formation of the blastula, and primitive streak, followed by organogenesis when individual organs develop within the newly formed germ layers.

Following gastrulation, cells are organized into sheets of connected cells (e.g., epithelial) or mesh of isolated cells (i.e., *mesenchyme*).

Each germ layer gives rise to specific tissues and organs in the developing embryo.

Ectoderm gives rise to the epidermis and other tissues that will form the nervous system.

Mesoderm is between the ectoderm and the endoderm.

Mesoderm gives rise to somites that form muscle, cartilage of the ribs and vertebrae, dermis, notochord, blood and blood vessels, bone, and connective tissue.

Endoderm gives rise to epithelium of the respiratory system, digestive system, and organs associated with the digestive system (e.g., liver and pancreas).

Gastrulation occurs when a blastula of one layer folds inward and enlarges to create the *three primary germ layers* (i.e., endoderm, mesoderm, and ectoderm).

Archenteron gives rise to the digestive tube.

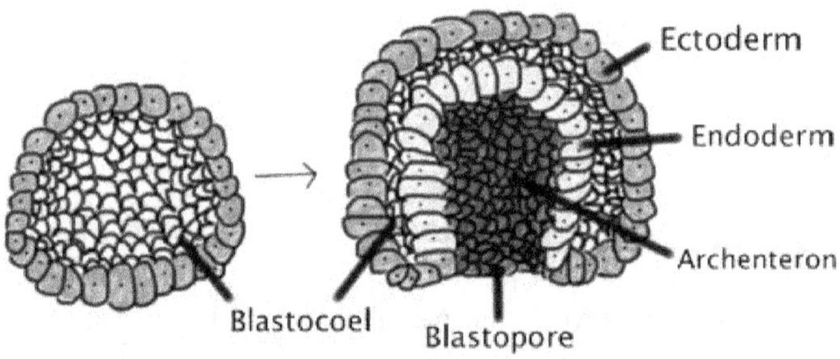

Blastula (left) becomes the gastrula (right) with the formation of the three primary germ layers

9. A is correct.

During the first *eight weeks* of development, myelination of the spinal cord does not occur.

Myelination of the *corticospinal tract* starts around 36 weeks and finishes around two years.

Most major developmental milestones before two years are due to myelination of nerve fibers.

10. E is correct.

Ectoderm germ layers give rise to the skin epidermis and the nervous system.

Endoderm germ layer gives rise to the lining of the digestive system, its associated glands, organs (e.g., liver and pancreas), and the lungs.

Mesoderm gives rise to most organs and systems, including excretory, reproductive, muscular, skeletal, and circulatory systems. If tissue is not specifically endoderm or ectoderm, it is likely to *be mesoderm derived.*

The incorrect choices are *mesoderm-derived* tissues or structures.

11. C is correct.

Fate mapping traces the embryonic origin of tissues in adults by establishing the correspondence between individual cells (or groups of cells) at one stage of development and their progeny at later stages.

When conducted at single-cell resolution, this process is *cell lineage tracing.*

Phylogenetic tree (i.e., an evolutionary tree) is a branching diagram showing the inferred evolutionary relationships (phylogeny) among species or other entities based upon similarities and differences in their physical or genetic characteristics.

Taxa joined in the tree are implied to have descended from a *common ancestor.*

Pedigree diagram shows the occurrence and appearance (i.e., *phenotypes*) of a gene in ancestors.

A pedigree presents family information in an easily readable chart.

Pedigrees use *standardized symbols*: *squares* represent males and *circles* females.

Pedigree construction is *family history*; details about an earlier generation may be uncertain as memories fade.

For unknown sex, a diamond is used.

Phenotype in question is represented by a filled-in (darker) symbol.

Heterozygotes are indicated by a shaded dot inside a symbol or a half-filled symbol.

Linkage map is a genetic diagram of an experimental population that shows the relative position of its known genes or genetic markers based on recombination frequency rather than the physical distance on chromosomes.

Genetic linkage is the tendency of genes proximal on a chromosome to be inherited together during meiosis.

Genes whose loci (i.e., position) are closer are *less likely* to be separated onto different chromatids during chromosomal crossover (i.e., during prophase I) and are genetically *linked.*

12. A is correct.

Hyaline cartilage tissue is the precursor of long bones in the embryo.

Fibrous cartilage is in intervertebral discs.

Dense fibrous connective tissue generally forms *tendons* and *ligaments.*

Elastic cartilage provides internal support and recoil to the external ear and epiglottis.

13. D is correct.

Gene expression changes as development proceeds; proteins are encoded by expressed genes.

During development, cells change their ability to respond to signals from tissues and induce changes in other cells. These changes during development are inherited by daughter cells.

Microtubules are involved in mitosis but do not change during development.

Somatic cells have the same genes (exceptions include gametes and B and T cells).

14. C is correct.

Acrosomes are at the tip of the sperm and contain specialized secretory molecules.

Acrosomal reaction is a signaling cascade involving the glycoproteins on the egg's surface.

Acrosin digests the zona pellucida and membrane of the oocyte.

Part of the sperm cell membrane fuses with the egg cell's membrane, and the contents of the head enter the egg fused with the plasma membrane. This allows the sperm to release its degradation enzymes, penetrate the egg's tough coating, and bind and fuse.

Pronucleus is the nucleus of a sperm or an egg cell after the sperm enters the ovum but before they fuse.

15. E is correct.

Three primary germ layers form during gastrulation in embryogenesis: *ectoderm*, *mesoderm,* and *endoderm.*

Ectoderm is the external germ layer giving rise to the skin, fingernails, and nervous system (including the eye).

Mesoderm is the middle germ layer and gives rise to most organ systems, including the musculoskeletal, cardiovascular, reproductive, and excretory systems.

Endoderm is the innermost germ layer and gives rise to the gallbladder, liver, pancreas, and epithelial lining of luminal structures and accessory digestive organs.

16. D is correct.

Ductus arteriosus is a blood vessel connecting the pulmonary artery to the proximal descending aorta in the developing fetus. It allows most blood from the right ventricle to bypass the fetus's fluid-filled non-functioning lungs. Upon closure at birth, the *ductus arteriosus* becomes the *ligamentum arteriosum.*

Ligamentum teres (i.e., round ligament) refers to several structures:

ligamentum teres *uteri* (i.e., round ligament of the uterus)

ligamentum teres *hepatis* (round ligament of the liver)

ligamentum teres *femoris* (i.e., ligament of the head of the femur)

17. C is correct.

Blastula invagination forms *mesoderm* as the third primary germ layer during gastrulation.

18. E is correct.

Homeotic genes encode homeodomain proteins involved in developmental patterns and sequences.

Homeodomain is a protein structural domain that binds DNA or RNA and is often transcription factors.

Homeodomain protein folding consists of a 60-amino acid helix-turn-helix structure in which short loop regions connect three alpha-helices. N-terminus of the two helices is antiparallel, and the longer C-terminal helix is perpendicular to the axes established by the first two.

*Third heli*x functions as a *transcription factor and interacts directly with DNA*.

19. D is correct.

Neurulation and *organogenesis* follow gastrulation.

Mitosis continues throughout development, but *cleavage* is a specific term reserved for *the first few cell divisions* when the *zygote* becomes the *morula*.

During *cleavage*, no growth occurs, and the morula is the same approximate size as the zygote.

Blastula formation precedes gastrulation.

Blastocoel is the fluid-filled central region of a blastula and forms early after fertilization when the zygote divides into many cells.

E: once *differentiated* (i.e., morphological, and biochemical distinct), a cell does not reverse this differentiation process unless it is a cancerous cell.

20. A is correct.

Trophoblast cells give rise to the outer layer of a blastocyst, provide nutrients to the embryo and develop into a large part of the placenta.

Trophoblasts form during the first stage of pregnancy and are the first cells differentiating from the fertilized egg.

21. B is correct.

Placenta connects the developing fetus to the uterine wall and allows nutrient uptake, waste elimination, and gas exchange via the mother's blood supply.

Umbilical cord connects the developing embryo (or fetus) and the placenta.

During prenatal development, the *umbilical cord* is physiologically and genetically part of the fetus and contains two arteries (the *umbilical arteries*) and one vein (the *umbilical vein*).

Umbilical veins supply the fetus with oxygenated, nutrient-rich blood from the placenta.

Fetal heart pumps deoxygenated, nutrient-depleted blood through the umbilical *arteries* to the placenta.

22. E is correct.

For vertebrates, *induction* is how a group of cells cause *differentiation* in another group of cells.

For example, cells that form the notochord induce the formation of the neural tube.

For example, induction in vertebrate development is eye formation, where the optic vesicles induce the ectoderm to thicken and form the lens placode (i.e., thickened portion of ectoderm becoming lens), which induces the optic vesicle to form the optic cup, which induces the lens placode to form the cornea.

Neural tube does develop into the nervous system but is not induced by another group of cells or tissue.

Thyroxin stimulating hormone (TSH) stimulates thyroxine secretion but is not induction.

Neurons synapse with other neurons via neurotransmitters (i.e., chemical messengers), but they do not induce changes in other tissues, as does induction.

23. D is correct.

Acrosomal reaction by the sperm is hydrolytic enzymes degrading the plasma membrane.

Acrosome is at the tip of sperm and contains specialized secretory molecules.

Acrosomal reaction is due to a signaling cascade involving the glycoproteins on the egg's surface.

Acrosin digests the zona pellucida and membrane of the oocyte, and the sperm releases its degradation enzymes to penetrate the egg's tough coating and allow the sperm to bind and fuse with the egg.

24. C is correct.

Endoderm is the innermost germ layer and gives rise to the inner lining of the respiratory and digestive tracts and associated organs.

Blastula refers to a hollow ball of embryonic cells arising from the morula. The blastula is *not* a germ layer.

Ectoderm is the outermost germ layer and gives rise to the hair, nails, eyes, skin, and central nervous system.

25. E is correct.

Inner cell mass is the cells inside the primordial embryo that forms before implantation and gives rise to the definitive structures of the fetus.

Primitive streak is a structure that forms in the blastula during the initial stages of embryonic development.

Primitive streak establishes:

> *bilateral symmetry*
>
> determines the *site of gastrulation*
>
> *initiates germ layer formation*

26. D is correct.

Fetal circulatory system changes as newborns begin using their lungs at birth.

Resistance in the pulmonary blood vessels decreases, causing an increase in blood flow to the lungs.

Umbilical blood flow ceases at birth, and blood pressure in the inferior vena cava decreases, which causes a decrease in pressure in the right atrium.

In contrast, the left atrial pressure increases due to increased blood flow from the lungs.

Increased left atrial pressure, coupled with decreased right atrial pressure, causes closure of the foramen ovale.

> *Ductus arteriosus* (i.e., connects the pulmonary artery to the aorta) constricts and subsequently is sealed.
>
> *Ductus venosus*, which shunts blood from the left umbilical vein directly to the inferior vena cava to allow oxygenated blood from the placenta to bypass the liver, completely closes within three months after birth.

Fetus produces adult hemoglobin a few weeks before birth (2 α and 2 β chains; alpha and beta) though lower amounts of fetal hemoglobin continue until the production completely stops.

After the first year, low fetal hemoglobin levels (2 α and 2 γ chains; alpha and gamma) are in the infant's blood.

27. B is correct.

Homeotic genes encode the related homeodomain proteins involved in developmental patterns and sequences.

Homeobox is a stretch of DNA about 180 nucleotides long that encodes a homeodomain (i.e., protein) in vertebrates and invertebrates.

Exons (expressed sequences) are retained during RNA processing of the primary transcript (hnRNA) into mRNA for translation into proteins.

28. A is correct.

Somatic cells have the same genome, but individual cells express different genes.

Differences in expression are *spatial* (i.e., cell type) or *temporal* (i.e., stages of development).

Ectoderm tissue arises after gastrulation as a primary germ layer.

Each germ layer retains its *determination*.

Somatic cells have identical genomes compared to their parents.

Gametes (via segregation and recombination during prophase I) have unique 1N genomes compared to their parents.

29. C is correct.

Yolk sac in humans gives rise to *blood cells* and *gamete-forming cells.*

Chorion is a double-layered membrane formed by trophoblast and extra-embryonic mesoderm.

Chorion gives rise to the fetal part of the placenta.

Luteal placental shift (7-9 weeks) is when the placenta develops enough to produce hormones to sustain the pregnancy.

Before this shift, the *placenta secretes progesterone* instead of the *corpus luteum.*

In humans, waste goes to the placenta and is received by the mother.

Non-placental organisms use allantois to collect waste.

30. A is correct.

Trophoblast is primarily responsible for forming placental tissue.

Trophoblast gives rise to the outer layer of a blastocyst, provides nutrients to the embryo, and develops into a large part of the placenta.

Trophoblasts form during the first stage of pregnancy and the first cells to *differentiate* from the fertilized egg.

31. C is correct.

Spina bifida is a bone abnormality that arises from the embryonic mesoderm germ layer.

A lesion to the mesoderm would affect the development of other structures based on different connective tissue (e.g., blood, blood vessels, muscles, and connective tissue of organs).

Thus, this lesion affects the development of blood vessels and muscles.

I: *intestinal epithelium* develops from *endoderm.*

II: skin, hair, and the nervous system develop from *ectoderm.*

32. B is correct.

Homeotic genes encode the related homeodomain proteins involved in *developmental patterns and sequences*.

Homeotic genes are involved in developmental patterns and sequences.

For example, homeotic genes determine where, when, and how body segments develop.

Alterations in these genes cause changes in body parts and structure patterns, sometimes resulting in dramatic effects.

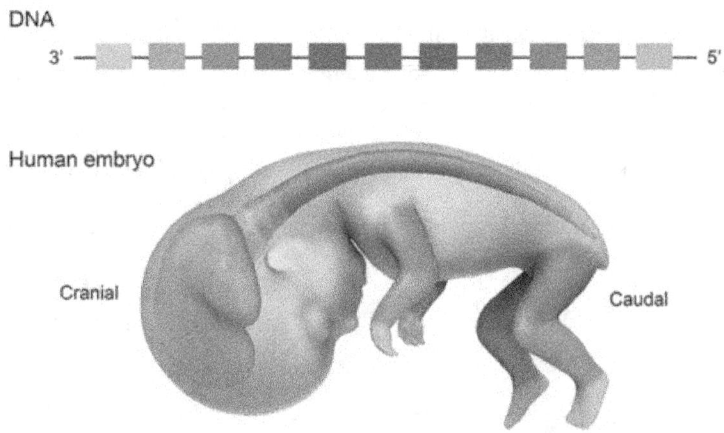

HOX genes encode homeodomain proteins as developmental regulators of the anterior-posterior axis. The linear sequence of genes on the chromosome corresponds to the cranial-caudal orientation

Homeodomain is a protein structural domain that binds DNA or RNA and is a transcription factor (i.e., facilitates polymerase binding and activation).

33. D is correct.

Mesoderm forms muscles, blood, bone, reproductive organs, and kidneys.

34. C is correct.

Primitive streak is a transitional structure formed at the onset of gastrulation.

Inner cell mass is converted into the trilaminar embryonic disc comprised of the three germ layers: ectoderm, mesoderm, and endoderm.

Primitive streak formation is not part of implantation but an early process of embryonic development.

35. A is correct.

During labor, the *oxytocin* hormone stimulates contractions of uterine smooth muscle.

36. D is correct.

Ectoderm cells are determined but not differentiated because they have the potential to develop into more than one type of tissue, but not any type.

Experimentally, ectoderm development was influenced (i.e., induced) by changing locations within the developing embryo. The underlying *mesoderm* differentiated because it released the molecular inducers as signals to the overlying ectoderm (undifferentiated).

Ectoderm cell *location* is essential because their development (i.e., differentiation into specific tissue/structures) is induced by the underlying mesoderm cells that send chemical substances (i.e., inducers) specific to the position of the mesoderm.

Mesoderm, not ectoderm, determines cellular differentiation at this stage of development.

Differentiated fate of the cells has not yet been determined because the transplanted cells would develop into wing feathers instead of claws.

Ectoderm cells cannot develop into any tissue because location influences the development of transplanted ectoderm cells.

37. D is correct.

Recognition of the sperm by the *vitelline envelope* (i.e., a membrane between the outer zona pellucid and inner plasma membrane) triggers the cortical reaction, transforming it into the hard *fertilization membrane* as a physical barrier to other spermatozoa (i.e., *slow block to polyspermy*).

Cortical reaction within the egg is analogous to the acrosomal reaction within the sperm.

38. C is correct.

Polyspermy in humans results in a nonviable zygote.

39. E is correct.

Mesoderm germ layer gives rise to many tissues.

Intestinal mucosa is derived from *endoderm*

Nerve is derived from *ectoderm*.

Lung epithelium is derived from *endoderm*.

40. C is correct.

Genomic imprinting is *epigenetic* (i.e., *heritable change*) when specific genes are expressed in a parent-of-origin-specific manner.

Genomic imprinting is an *epigenetic process* involving *DNA methylation* and *histone modulation* to achieve single-allelic gene expression without altering the original genetic sequence.

41. C is correct.

Embryos require *greater protein translation rates* using ribosomes to read mRNA transcripts.

42. E is correct.

Zygote (i.e., fertilized egg) is the first structure to form during human fertilization that undergoes rapid cell divisions without significant cellular growth to produce a cluster of cells of the same size as the original zygote.

Cells derived from cleavage are *blastomeres* and form a solid mass known as a *morula.*

Cleavage ends with the formation of the *blastula.*

> Ovulation → fertilization (sperm and oocyte) → diploid zygote (undergoes cleavage) → morula (solid ball of cells) → blastocoel (undergoes invagination) → blastopore (ectoderm and endoderm)

43. A is correct.

Capacitation is the final step in spermatozoa maturation. It is required for competence to fertilize oocytes.

Capacitation destabilizes the acrosomal sperm head membrane, allowing greater binding between sperm and oocyte by removing steroids (e.g., cholesterol) and non-covalently bound glycoproteins, which increases membrane fluidity and Ca^{2+}permeability.

Ca^{2+} influx produces intracellular cAMP levels and increased sperm motility.

44. C is correct.

Endoderm develops into *epithelial linings* of the digestive and respiratory tracts, parts of the liver, thyroid, pancreas, and bladder lining.

45. E is correct.

Homeotic genes influence the development of specific structures in plants and animals, such as the Hox and ParaHox genes, which are essential for segmentation.

Homeotic genes determine where, when, and how body segments develop in organisms.

Alterations in these genes cause changes in patterns of body parts, sometimes causing dramatic effects, such as legs growing in place of antennae or an extra set of wings.

Loss-of-function mutations result in the gene product with less or no function.

An allele with loss of function (i.e., null allele) is an *amorphic* (i.e., complete loss of gene function) mutation.

Phenotypes associated with such mutations are often recessive.

Exceptions are haploid or *haploinsufficiency* (i.e., a diploid organism has a single copy of the functional gene) when the reduced dosage of a normal gene product is not enough for a typical phenotype.

46. A is correct.

Cells may assume several fates and are not yet terminally *differentiated.*

Gastrula cells influenced by their surroundings are *competent.*

Ectoderm layer gives rise to the eye, among other structures.

Cells can become other ectoderm tissue (e.g., gills).

47. B is correct.

Proteases and *acrosin* enzymes degrade the protective barriers around the egg and allow the sperm to penetrate.

Acrosomal reaction by the sperm is hydrolytic enzymes degrading the plasma membrane.

Acrosomes are at the tip of the sperm and contain specialized secretory molecules.

Acrosomal reaction is due to a signaling cascade involving the glycoproteins on the egg's surface.

Acrosin digests the zona pellucida and membrane of the oocyte, and the sperm releases its degradation enzymes to penetrate the egg's tough coating and allow the sperm to bind and fuse with the egg.

48. A is correct.

Human blastocyst implants in the uterine wall about a *week* after fertilization.

49. B is correct.

Ectoderm develops into the nervous system, the epidermis, the eye lens, and the inner ear.

Endoderm develops into the lining of the digestive tract, lungs, liver, and pancreas.

Mesoderm develops into the connective tissue, muscles, skeleton, circulatory system, gonads, and kidneys.

50. A is correct.

After the infant's first breath, the newborn's cardiovascular system constricts the *ductus arteriosus* (i.e., connects the pulmonary artery to the aorta) and converts it to the *ligamentum arteriosum.*

Fetal circulatory systems change at birth as the newborn uses its lungs.

Resistance in pulmonary blood vessels decreases, increasing blood flow to the lungs.

At birth, umbilical blood flow ceases, and blood pressure in the inferior vena cava decreases, which causes a decrease in pressure in the right atrium.

In contrast, the left atrial pressure increases due to increased blood flow from the lungs.

Increased left atrial and decreased right atrial pressure causes *closure* of the *foramen ovale.*

Ductus venosus, which shunts blood from the left umbilical vein directly to the inferior vena cava to allow oxygenated blood from the placenta to bypass the liver, completely closes within three months after birth.

Diagram below

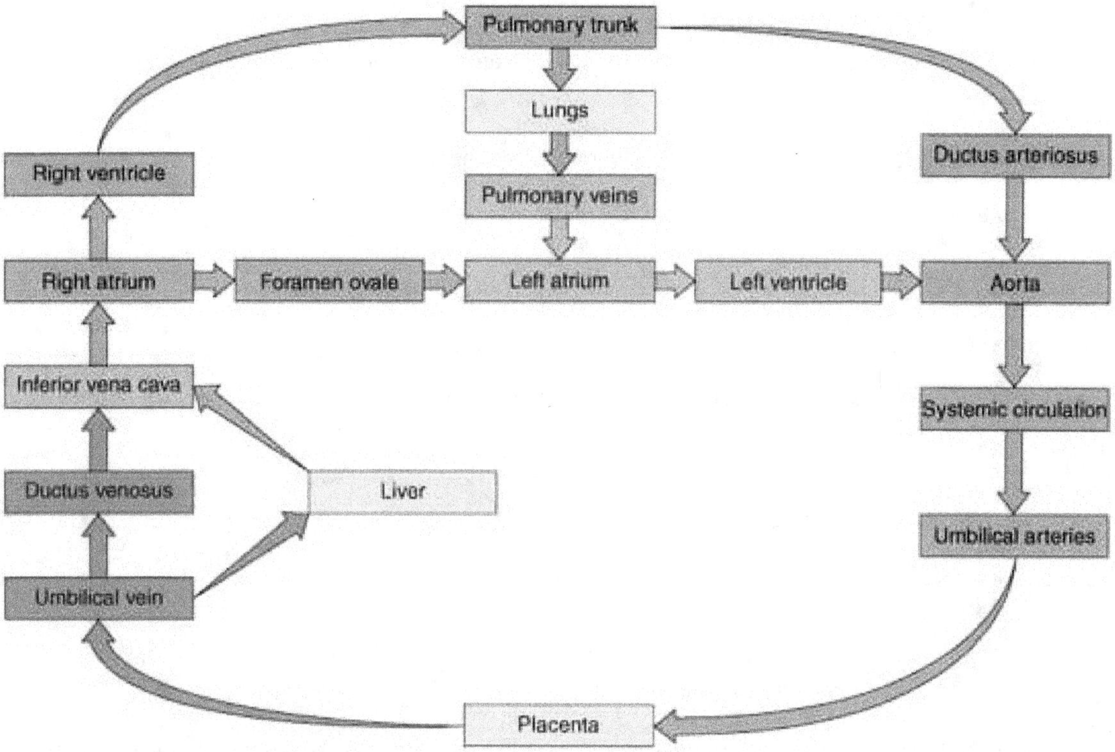

Adult and fetal circulatory systems with ductus arteriosus and ductus venosus provide shunt pathways

51. C is correct.

Genomic imprinting is an *epigenetic* (i.e., heritable changes in gene activity) process involving DNA methylation and histone remodeling for monoallelic (i.e., single allele) gene expression without altering the genetic sequence within the genome.

Epigenetic changes are established in the germline (i.e., cells that give rise to gametes of egg/sperm) and can be maintained through mitotic divisions.

52. E is correct.

Amniotic fluid, surrounded by the amnion, is a liquid environment around the egg that protects it from shock.

Embryonic membranes include:

1) ***chorion*** lines the inside of the shell and permits gas exchange;

2) ***allantois*** is a saclike structure developed from the digestive tract and functions in respiration, excretion, and gas exchange with the external environment;

3) ***amnion*** encloses *amniotic fluid* as a watery environment for embryogenesis and shock protection;

4) ***yolk sac*** encloses the yolk and transfers food to the developing embryo.

53. A is correct.

Mesoderm gives rise to the *entire circulatory system, muscle*, and most other tissue between the *gut and skin* (excluding the nervous system).

Ectoderm gives rise to skin, the nervous system, the retina, and the lens.

Differentiated endoderm has a restricted fate that does not include heart tissue.

Endoderm gives rise to the *inner lining of the gut.*

54. C is correct.

At the end of the *first trimester*, an ultrasound can determine the sex of the fetus.

55. E is correct.

Eggs are viable and can be fertilized 12-24 hours after ovulation.

56. B is correct.

Blastulation begins when the *morula* develops a fluid-filled cavity (i.e., blastocoel). By the fourth day, it becomes a hollow sphere of cells (i.e., blastula).

Gastrula is the embryonic stage characterized by the three primary germ layers (i.e., endoderm, ectoderm, and mesoderm), the blastocoel, and the archenteron.

Early gastrula is two-layered (i.e., ectoderm and endoderm); shortly afterward, a third layer (i.e., mesoderm) develops.

Gastrulation is followed by organogenesis, when individual organs develop within newly formed germ layers.

Morula is the solid ball of cells from the early cleavage stages in the zygote.

Zygote is the (2N) cell formed by the fusion of two (1N) gametes (e.g., ovum and sperm).

57. B is correct.

Proteases and *acrosin* enzymes degrade the protective barriers around the egg and allow the sperm to penetrate.

Acrosomal reaction by the sperm is hydrolytic enzymes degrading the plasma membrane.

Acrosomes are at the tip of the sperm and contain specialized secretory molecules.

Acrosomal reaction is due to a signaling cascade involving the glycoproteins on the egg's surface.

Acrosin digests the zona pellucida and membrane of the oocyte, and the sperm releases its degradation enzymes to penetrate the egg's tough coating and allow the sperm to bind and fuse with the egg.

58. C is correct.

Regarding fertilization, the *vagina's acidic environment* destroys millions of sperm cells.

59. B is correct.

Zygote is a fertilized egg undergoing rapid cell divisions without significant cellular growth to produce a cluster of cells of the same total size as the original zygote.

Cells derived from cleavage are *blastomeres* and form a solid mass known as a *morula.*

Cleavage ends with the formation of the *blastula.*

Among species, depending mainly on the amount of yolk in the egg, cleavage is:

holoblastic (i.e., total or entire cleavage) or

meroblastic (i.e., partial cleavage).

Egg pole with the highest yolk concentration is the *vegetal pole*, while the opposite is the *animal pole.*

Humans undergo *holoblastic cleavage* (i.e., total cleavage).

60. B is correct.

Fertilized egg is a *zygote* undergoing rapid cell divisions without significant growth, producing a cluster of cells of the same total size as the original zygote.

Cells derived from zygote cleavage are *blastomeres* and form a solid mass known as a *morula.*

Cleavage ends with the formation of the *blastula* (i.e., hollow ball of cells).

In different species, mainly depending on the amount of yolk in the egg, cleavage is *holoblastic* (i.e., total or entire cleavage) or *meroblastic* (i.e., partial cleavage).

Egg pole with the highest yolk concentration is the *vegetal pole*, while the opposite end is the *animal pole.*

Humans undergo *holoblastic cleavage.*

Notes for active learning

Notes for active learning

Notes for active learning

DNA and Protein Synthesis – Detailed Explanations

1. D is correct.

Histones are basic (i.e., positively charged) proteins associated with nuclei DNA to condense chromatin.

Nuclear DNA does not appear in free linear strands; instead, it is highly condensed and wrapped around histones (i.e., positively-charged proteins) to fit inside the nucleus and form chromosomes.

Three major types of RNA engage in gene expression:

1) *messenger* RNA (mRNA) molecules carry the coding sequences (i.e., "blueprints") for protein synthesis and are transcripts;

2) *ribosomal* RNA (rRNA) forms the core of a cell's ribosomes (i.e., macromolecular cellular particles where protein synthesis takes place);

3) *transfer* RNA (tRNA) molecules transport amino acids (i.e., protein building blocks) to the ribosomes during protein synthesis.

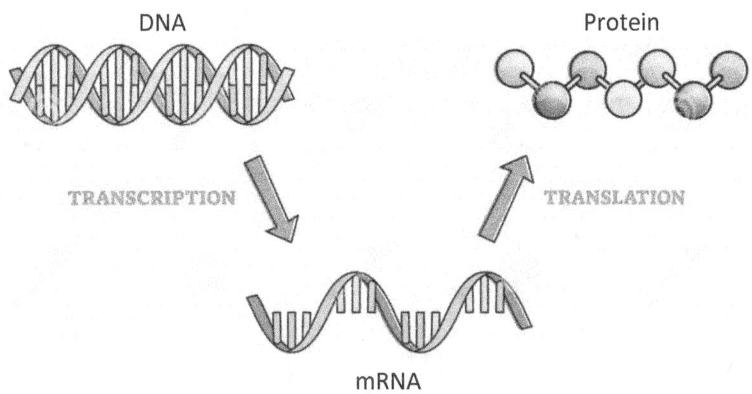

Gene expression with DNA transcribed into mRNA, which is translated into protein

2. D is correct.

Frederick Griffith (1928) reported an early experiment suggesting that bacteria transfer genetic information through *transformation.*

Griffith observed a *transforming principle*, where heat-killed S (smooth) bacteria was destroyed, and (now known) its DNA survived the process and was incorporated by R (rough) strain bacteria.

S genetic fragments protected R strain bacteria from host immunity and killed the host.

3. E is correct.

Depurination is the hydrolysis of the glycosidic bond of DNA and RNA purine nucleotides guanine or adenine.

Nucleotides have sugar, phosphate, and base (A, C, G, T or U).

For DNA, the sugar is deoxyribose, while RNA contains ribose.

Glycosidic bonds are hemiacetal groups of a saccharide (or an amino acid) and the hydroxyl group of an organic compound (e.g., alcohol).

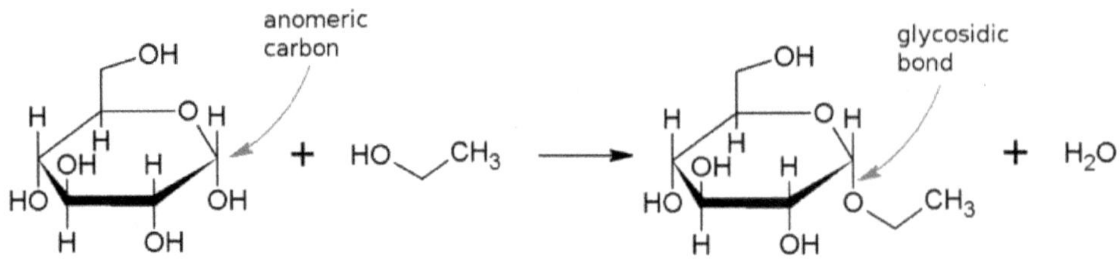

Glycosidic bond formation from the nucleophilic attack of alcohol on the anomeric carbon of a hemiacetal

4. A is correct.

Unequal crossing over is a gene duplication (or deletion) event that deletes a sequence in one strand of the chromatid. It replaces it with duplication from its sister chromatid in mitosis, or homologous chromosomes recombine during prophase I in meiosis.

5. D is correct.

Sulfur is in the amino acid cysteine but is absent in nucleic acids.

Hershey-Chase (i.e., associate *blender*) experiment in 1952 used radiolabeled molecules of phosphorus (^{32} P for nucleic acids) and sulfur (^{35}S for proteins) to determine whether nucleic acids (phosphorus) or protein (sulfur) carried the genetic information.

Nucleic acids contain C, H, O, N and P and are polymers of *nucleotide* subunits.

Nucleic acids (e.g., DNA, RNA) encode cellular information for protein synthesis and replication.

6. D is correct.

mRNA is the subclass of RNAs molecule translated into proteins.

7. A is correct.

Aging of normal cells is associated with a loss of *telomerase activity*.

8. C is correct.

Translation is initiated on ribosomes within the cytoplasm.

A: *transcription* (not translation) occurs in the nucleus.

B: *Golgi* receives proteins from the rough endoplasmic reticulum for processing and sorting.

Golgi modifies (e.g., adds carbohydrate groups) and sorts proteins in the *secretory pathway*.

D: proteins destined for the *lumen of the rough endoplasmic reticulum* have at their amino terminus a particular sequence of amino acids referred to as a *leader sequence* (about 6 to 10 amino acids).

Signal recognition protein (SRP) recognizes the leader sequence and binds a receptor on the rough ER, attaching the ribosome and the nascent polypeptide to the endoplasmic reticulum (ER) membrane.

9. E is correct.

When a gene is duplicated by crossing over between chromatids, the gene on the other chromatid is deleted.

Unequal crossing over is a gene duplication (or deletion) event that deletes a sequence in one strand of the chromatid. It replaces it with duplication from its sister chromatid in mitosis, or homologous chromosomes recombine during prophase I in meiosis.

10. C is correct.

To show that DNA, not RNA, protein, or other cell components, was responsible for transformation, Avery, MacLeod, and McCarty (1944) used several biochemical tests.

Trypsin, chymotrypsin, and ribonuclease (enzymes digesting proteins or RNA) did not affect the transforming agent causing the disease.

DNase treatment degrades DNA and destroys the extract's ability to cause disease.

Streptococcus pneumoniae (i.e., pneumococcus) is a Gram-positive pathogenic bacterium.

S. pneumoniae was recognized as a major cause of pneumonia in the late 19th century and has been the subject of numerous humoral immunity (i.e., antibody-mediated) studies.

11. B is correct.

Western blotting separates proteins by electrophoresis and is commonly used to identify the presence of HIV antibodies (proteins).

Blotting techniques rely on gel electrophoresis to separate DNA, RNA, or proteins based on size.

After resolution (separation) by electrophoresis, the gel (containing resolved products) is blotted (i.e., transferred to nitrocellulose).

After transferring the macromolecules (DNA, RNA, or proteins) from the gel to the blotting paper by capillary action, the blotting paper is probed by specific markers that hybridize with complementary sequences fixed on the blotting paper.

A: *Eastern blotting* does not exist.

C: *Northern blotting* uses RNA in gel electrophoresis.

D: *Southern blotting* uses DNA in gel electrophoresis.

12. C is correct.

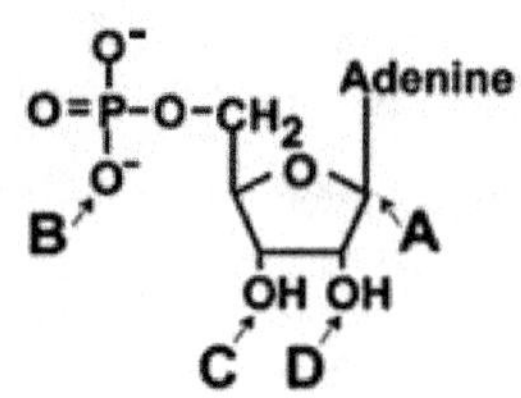

Position C represents the ribose sugar's 3' hydroxyl (~OH).

As in DNA, *the 3' hydroxyl is the site of attachment.*

RNA and DNA require a free 3' OH for the nucleic acid to increase length.

Position D contains a 2' hydroxyl (~OH) that distinguishes this sugar as ribose, as opposed to DNA which would lack the 2' hydroxyl (deoxyribose = without oxygen) at the 2' position of the sugar.

13. E is correct.

Adenine and guanine are *purines*.

C<u>y</u>tosine and thymine/uracil are *pyrimidines*; note the presence of *y* for pyrimidines.

A: DNA strands are *antiparallel*: one strand has a 5'→ 3' polarity, and its complementary strand 3' → 5'.

B: DNA consists of nucleotides, which have a phosphate group, a deoxyribose sugar, and a base (A, C, G, T).

D: cytosine binds guanine with three hydrogen bonds.

Adenine binds thymine (in DNA) or uracil (in RNA) with two hydrogen bonds.

14. C is correct.

Tumor suppressor gene protects a cell from aberrant cell cycles.

When this gene mutates to cause a loss (or reduction) in its function, the cell can progress to cancer, usually with other genetic changes.

The loss of *tumor suppressor genes* may be more important than *proto-oncogene* activation for forming many types of human cancer cells.

Apoptosis is the process of programmed cell death (PCD) that may occur in multicellular organisms.

Biochemical events lead to characteristic cell changes (morphology) and death, including blebbing, cell shrinkage, nuclear fragmentation, chromatin condensation, and chromosomal DNA fragmentation.

In contrast to necrosis (i.e., traumatic cell death that results from acute cellular injury), apoptosis confers advantages during an organism's lifecycle.

For example, a human embryo separates fingers and toes because cells between the digits undergo *apoptosis*.

Unlike necrosis, apoptosis produces cell fragments called *apoptotic bodies* that phagocytic cells can engulf and quickly remove before the cell contents spill onto surrounding cells and cause damage.

A: *telomerase* is an enzyme that adds DNA sequence repeats (i.e., TTAGGG) to the 3' end of DNA strands in the telomere regions at the ends of eukaryotic chromosomes.

This region of repeated nucleotides as telomeres with noncoding DNA hinders the loss of essential DNA from chromosome ends. When the chromosome is copied, 100–200 nucleotides are lost, causing no damage to the coding region of the DNA.

Telomerase is a reverse transcriptase that carries its RNA molecule used as a template when it elongates telomeres that have been shortened after each replication cycle.

Embryonic stem cells express telomerase, allowing them to divide repeatedly.

In adults, telomerase is highly expressed in cells that divide regularly (e.g., male germ cells, lymphocytes, and specific adult stem cells). Telomerase is not expressed in most adult somatic cells.

15. B is correct.

Aminoacyl tRNA synthetase (enzyme) uses energy from ATP to attach a specific amino acid to tRNA.

16. C is correct.

DNA was the transforming principle verified in experiments by Avery, MacLeod, and McCarty (1944) and Hershey and Chase (1952).

17. D is correct.

There are 4 different nucleotides (adenine, cytosine, guanine, and thymine/uracil).

Each codon is composed of 3 nucleotides.

Therefore, there must be 64 (4^3) possible variations of codons to encode for the 20 amino acids.

Genetic code is *degenerate* (i.e., redundant) because several codons encode for the same amino acid.

61 codons encode for amino acids and 3 stop codons that terminate translation.

18. E is correct.

Signal sequence at the N-terminus of the polypeptide targets proteins to organelles (e.g., chloroplast, mitochondrion).

19. C is correct.

Adenine pairs with thymine *via* 2 hydrogen bonds, while guanine pairs with cytosine *via* 3 hydrogen bonds.

Treatment of DNA with 2-aminopurine causes the adenine-thymine (A-T) base pair to be replaced with a guanine-thymine (G-T) base pair (before replication).

Thymine is replaced by cytosine: G-C base pair after replication.

This single-point mutation is incorporated into future generations.

If the mutation had been corrected before replication (via proofreading mechanisms during replication), there would be no change in the DNA base sequence.

20. D is correct.

Codons with two bases would be insufficient because the four bases in a two-base codon would form $4^2 = 16$ pairs, less than the 20 combinations needed to specify the amino acids.

A triplet is sufficient because four bases in a three-base codon can form $4^3 = 64$ pairs, enough to encode the 20 amino acids.

21. B is correct.

After one replication, the DNA was at an intermediate density between ^{14}N and ^{15}N.

After two replications, there were two densities – one band in the centrifuge tube was an intermediate between the ^{14}N and ^{15}N, while the other consisted of ^{14}N.

The *semiconservative* DNA replication model was one of three tested by the Meselson-Stahl (1958) experiment.

> ***Semiconservative replication*** would produce two copies containing one original and one new strand.
>
> ***Conservative replication*** would leave the two original template DNA strands in a double helix and produce a copy composed of two new strands containing the new DNA base pairs.
>
> ***Dispersive replication*** would produce two copies of the DNA, each containing distinct regions of DNA composed of either original or new strands.

In the Meselson-Stahl (1957-58) experiments, *E. coli* were grown for several generations in a medium with ^{15}N.

When DNA was extracted and separated by centrifugation, the DNA separated according to density.

The E. coli cells with ^{15}N in their DNA were transferred to a ^{14}N medium and divided.

The DNA of the cells grown in a ^{15}N medium had a higher density than cells grown in a standard ^{14}N medium.

Since *conservative replication* would result in equal amounts of DNA of the higher and lower densities (but no intermediate density), conservative replication was excluded.

This result was consistent with *semiconservative* and *dispersive replication.*

Semiconservative replication yields one double-stranded DNA with ^{15}N DNA and one with ^{14}N DNA.

Dispersive replication would result in double-stranded DNA, with both strands having mixtures of ^{15}N and ^{14}N DNA, which would have appeared as DNA of an *intermediate density*.

22. C is correct.

DNA microarray (or *DNA chip*) collects microscopic DNA spots attached to a solid surface. DNA microarrays measure the expression levels of many genes or genotype multiple regions of a genome.

Each DNA spot contains picomoles (10^{-12} moles) of a specific DNA sequence.

Short sections of a gene (or other DNA element) hybridize cDNA *probes*.

An array simultaneously uses tens of thousands of probes; microarrays evaluate many genetic tests in parallel.

23. B is correct.

Avery, MacLeod, and McCarty (1944) demonstrated that DNA was the *transforming principle* for nonvirulent strains of pneumococcus.

Enzymes that destroyed nucleic acids destroyed the transforming activity, strengthening their hypothesis.

24. A is correct.

Codon-anticodon hybridization (i.e., bonding interaction) occurs between mRNA (*codon*) and tRNA (*anticodon*) during translation for protein synthesis.

25. B is correct.

Northern blot is a molecular biology technique to study gene expression by detecting RNA expression levels. Northern blotting uses electrophoresis to separate RNA samples by size. It involves the capillary transfer of RNA from the electrophoresis gel to the blotting membrane. It uses a hybridization probe via complementary hydrogen bonding to target expressed RNA fragments (i.e., expressed genes).

Eukaryotic mRNA is isolated using oligo (dT) cellulose chromatography to hybridize mRNA with a poly-A tail. The sample is resolved (i.e., separated) by gel electrophoresis.

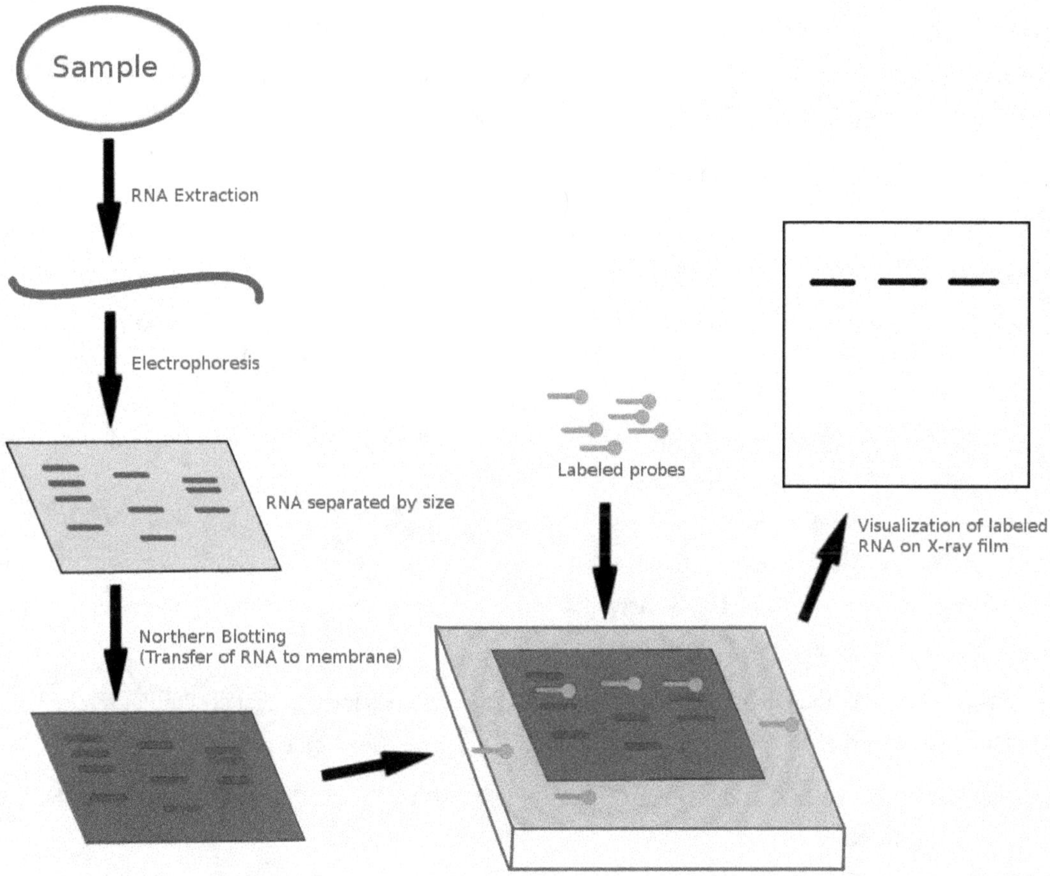

Northern blot uses RNA resolved by electrophoresis to evaluate gene expression using specific probes

Electrophoresis gels are fragile, and probes cannot enter the gel matrix. After resolution by electrophoresis, the size-separated RNA sample is transferred to a positively charged nylon (or nitrocellulose) membrane through capillary blotting. The negatively-charged mRNA adheres to the positive charge on the nylon.

In situ hybridization uses labeled probes of complementary DNA (cDNA) or RNA to localize a specific DNA or RNA sequence in a tissue section (*in situ*). Probes hybridize to the target sequence at an elevated temperature, and the excess probe is washed away.

26. E is correct.

Chromosomes replicate during the synthesis (S) phase of interphase.

27. E is correct.

DNA contains T, which is replaced with U in RNA.

28. B is correct.

Duplicated genes are closely related but diverged in sequence and function over evolutionary time.

29. D is correct.

Hershey and Chase showed in 1952 that when bacteriophages (i.e., viruses), composed of DNA and protein, infect bacteria, their DNA enters the host bacterial cell while their protein does not.

Hershey and Chase grew separate populations of viruses and incorporated radioactive sulfur (^{35}S to label protein) or phosphorus (^{32}P to label DNA) into the bacteriophages.

Two groups of viral progeny contained either ^{32}P or ^{35}S radioactive isotopes.

Separate aliquots of the labeled progeny were allowed to infect unlabeled bacteria. The viral ^{35}S protein coats remained outside the bacteria, while the ^{32} P DNA entered the bacteria.

Centrifugation separated the phage protein coats from the bacteria.

Bacteria were lysed to release the phages.

Hershey and Chase experiment demonstrated that the DNA, not protein, was the *transforming* molecule that entered the bacteria from a viral infection.

30. C is correct.

Replication forks open double-stranded DNA by disrupting hydrogen bonds between complementary nucleotide base pairs (e.g., A bonded to T and C bonded to G).

Gyrase cuts one strand of the DNA backbone and relaxes the positive supercoil that accumulates as helicase separates the two strands of DNA.

Ligase seals the backbone of DNA (i.e., joins Okazaki fragments) by forming phosphodiester bonds between deoxynucleotides in DNA.

31. A is correct.

Four atoms in the peptide bond are in the box.

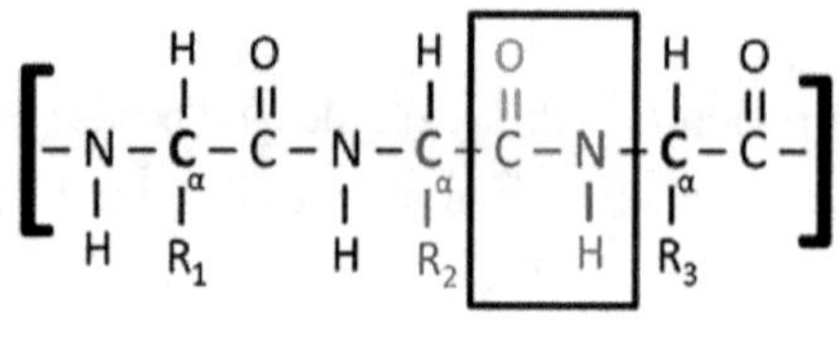

Note that the oxygen on the carbonyl is oriented 180° from the H (antiperiplanar) on the nitrogen because the lone pair on the nitrogen participates in resonance, and the peptide bond is rigid (i.e., double bond like character).

32. D is correct.

DNA and ribozymes of RNA (discovered in 1982) are capable of self-replication.

Protein functions include:

1) peptide hormones as chemical messengers transported within the blood,

2) enzymes that catalyze chemical reactions by lowering the energy of activation,

3) structural proteins for physical support within the cells, tissues, and organs,

4) transport proteins as carriers of important materials, and

5) immune system antibodies bind foreign particles (i.e., antigens).

33. B is correct.

Eukaryote RNA polymerase needs *transcription factors* (i.e., DNA binding proteins) to bind the promoter (on mRNA) and initiate basal-level transcription.

34. E is correct.

Percent of adenine cannot be determined because RNA is a single-stranded molecule.

Base-pairing rules for DNA (i.e., Chargaff's rule) are for double-stranded DNA but not single-stranded RNA.

35. A is correct.

I: AUG sequences are not only the initial start codon downstream of the initial *start codon* (AUG). If this sequence is not the start codon, this change could result in a stop codon (UAA).

II: *genetic code* is read 5′ to 3′ and AUG (encoding methionine) is a start codon.

A change in the start sequence in the mRNA (AUG to AAG) causes a failure in initiating translation.

III: changes in the first, second, or third amino acid may change U to A in the resultant amino acid.

The third position of the codon is the *wobble position* because the specified amino acid often does not change.

Nucleotide changes may not change the specified amino acid because the genetic code is redundant (i.e., most amino acids are encoded by more than one codon).

For example, AUU and AUA codons both encode isoleucine.

Genetic code is *degenerate*, whereby changing a nucleotide (often in the 3rd position) does not change the amino acid encoded by the triplet codon.

Each codon (three nucleotides) encodes for one amino acid, and there is no ambiguity in the genetic code.

36. B is correct.

Genetic code (mRNA into protein) has several amino acids specified by more than one codon, three stop codons, and each is 3 nucleotides (bases) long.

37. D is correct.

Polymerase I (Pol I) adds nucleotides at the RNA primer-template junction (i.e., the origin of replication) and is involved in excision repair with 3'-5' and 5'-3' exonuclease activity and processing of Okazaki fragments generated during lagging strand synthesis.

B: *primase* adds the first two RNA primers at the start of DNA replication because the DNA polymerase must bond to double-stranded molecules.

RNA primer is removed by DNA polymerase I after the newly synthesized DNA strand has been replicated via DNA polymerase III.

a: *template strand* of parental DNA

b: *leading strand* of newly synthesized DNA

c: *lagging strand* (Okazaki fragments) of newly synthesized DNA

d: *replication fork* with helicase opening and unwinding the double-stranded DNA

e: *RNA primer* synthesized by primase

f: *direction* of DNA strand synthesis

DNA replication

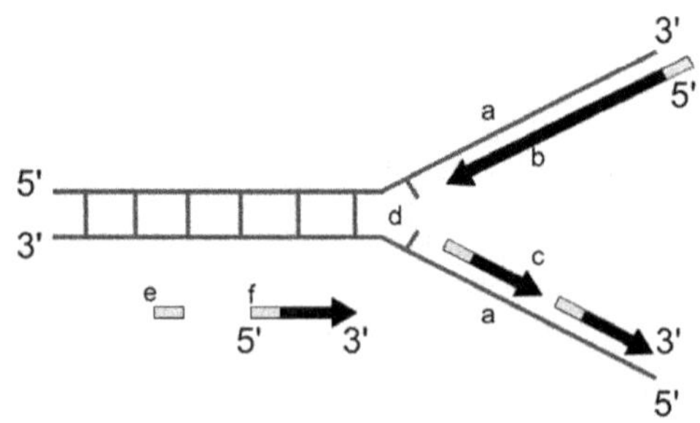

38. E is correct.

Okazaki fragments are associated with the lagging DNA strand and (like all DNA) are synthesized in a 5'→ 3' direction by DNA polymerase III.

A: DNA polymerase I removes the short sequence RNA primers deposited by primase needed for the anchoring of the polymerase III to DNA for the synthesis of DNA in a 5'→ 3' direction.

B: Okazaki fragments are used to replicate the lagging strand and are covalently linked by *DNA ligase* (not DNA polymerase I), forming a continuous DNA strand.

D: Okazaki fragments are not synthesized to fill in gaps after removing the RNA primer by DNA polymerase I.

Diagram below

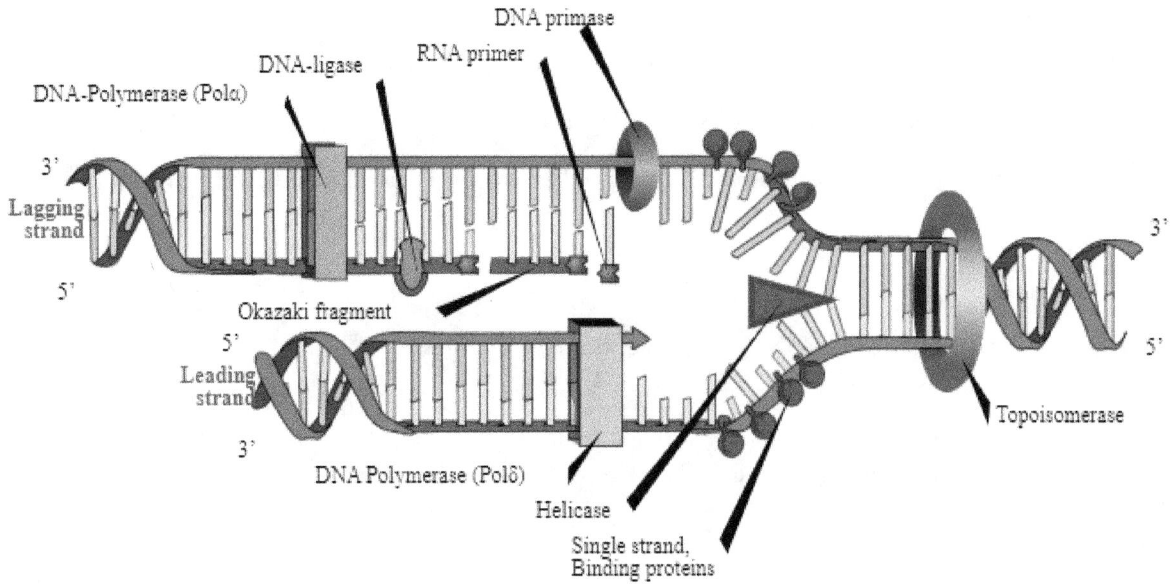

DNA semiconservative replication with associated proteins for leading and lagging strands. DNA polymerase proceeds from 5' to 3' along parental strands.

39. C is correct.

Ribosomal RNAs (rRNA) form a large subunit and a small subunit.

During translation, mRNA is between the small and large subunits, and the ribosome catalyzes the formation of a peptide bond between the two amino acids held by the rRNA.

A single mRNA can be translated simultaneously by multiple ribosomes.

Ribosomes catalyze the formation of a *peptide bond* between two amino acids tethered by rRNA.

Ribosomes have three binding sites: A, P, and E.

Peptidyl transferase catalyzes this reaction.

A site binds an aminoacyl-tRNA (i.e., a tRNA bound to an amino acid).

Amino (NH_2) group of the aminoacyl-tRNA, with the new amino acid, attacks the ester linkage of peptidyl-tRNA (in the P site), the last amino acid of the growing chain, forming a new peptide bond.

tRNA holding the last amino acid moves to the E site, and the aminoacyl-tRNA is now the peptidyl-tRNA.

40. A is correct.

Protein synthesis does require energy.

B: rRNA is part of the ribosome and is necessary for proper binding of the ribosome to mRNA.

C: tRNA brings an amino acid to the ribosome, where it interacts with the mRNA of the proper sequence.

D: tRNA does have the amino acid bound to its 3' end.

41. B is correct.

Polymerase chain reaction (PCR) requires the sequence at the ends of the fragment to be amplified.

From the ends, primers use complementary base pairing to anneal the target fragment and permit amplification.

No knowledge of the region between the ends is required because the parent strands will be the template used by the DNA polymerase.

42. D is correct.

Shape of tRNA is determined primarily by *intramolecular base pairing.*

43. E is correct.

In prokaryotic cells, methylated guanine contributes to correcting mismatched pairs of bases.

44. A is correct.

Magnesium is a *divalent* mineral that DNA and RNA polymerases use as a cofactor (i.e., catalyst not consumed in the reaction) to stabilize interactions between polymerase and negative charge on the nucleic acid backbone.

45. C is correct.

Ribosome structure is created by the internal base-pairing of rRNA and ribosomal proteins.

46. A is correct.

Three types of RNA are *mRNA*, *tRNA*, and *rRNA*, which DNA encodes.

rRNA is synthesized in the nucleolus within the nucleus.

tRNA functions as a carrier of amino acid molecules.

Unlike mRNA, tRNA is a comparatively short ribonucleotide polymer of RNA subunits.

Although tRNA is single-stranded, there are double-stranded segments where the nucleotide chain loops back (i.e., hairpin turns) with hydrogen bonds between complementary base pairs; like DNA, two hydrogen bonds between A and U and three hydrogen bonds between C and G.

mRNA is the template for protein synthesis and has a poly-A tail, which functions as a *molecular clock* for mRNA degradation.

47. D is correct.

Phosphate group is the chemical group at the 5' end of a single polynucleotide strand.

48. E is correct.

Puromycin is an analog with a similar shape to tRNA.

Puromycin joins the ribosome, forms one peptide bond, and becomes covalently attached to the nascent protein.

However, since it lacks a carboxyl group, it cannot be linked to the next amino acid, and protein synthesis terminates prematurely.

A: *initiation* requires binding a single aminoacyl-tRNA (the initiator) to the ribosome and is unaffected.

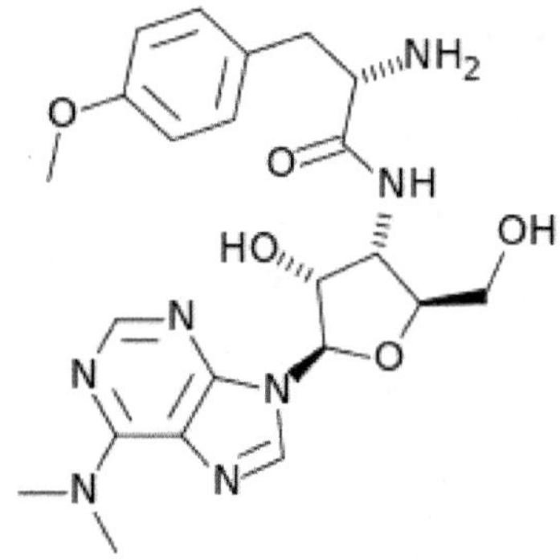

Puromycin

B: *aminoacyl-tRNA* enters the large subunit of the ribosome at the A site during elongation.

D: *puromycin* lacks a carboxyl group and can only form one bond, so peptide synthesis stops prematurely.

49. C is correct.

In *E. coli* cells, DNA polymerase I degrades the RNA primer portion of Okazaki fragments.

50. D is correct.

DNA polymerase I proofreading increases replication fidelity by monitoring for mismatched pairs originating from the high processivity of polymerase III that rapidly replicates DNA.

Bacteria have a much lower DNA replication rate of about 1 in 1,000, increasing the mutation rate of bacteria.

51. C is correct.

DNA is a nucleotide polymer of the deoxyribose sugar, a phosphate group, and a nitrogenous base (e.g., A, C, G, T). Phosphodiester bonds join the nucleotides in DNA's backbone.

52. A is correct.

In *E. coli* cells, DNA polymerase III synthesizes most of the Okazaki fragments.

53. E is correct.

All the molecules, except cysteine, are nitrogenous bases – the component molecules of DNA and RNA (e.g., mRNA, rRNA & tRNA).

Nitrogenous bases guanine (G) and adenine (A) are purines, while cytosine (C) and thymine (T is in DNA) or uracil (U is in RNA) are pyrimidines.

Cysteine is an amino acid (not a nitrogenous base), and amino acids are the monomers for proteins.

54. A is correct.

Restriction enzyme (*restriction endonuclease*) cut DNA at specific nucleotide sequences (i.e., restriction sites).

Restriction enzymes are a defense mechanism against invading viruses in bacteria and archaea.

Prokaryotes have restriction enzymes that selectively cleave foreign DNA.

Host DNA is protected by a modification enzyme (i.e., methylase) that alters the prokaryotic DNA and prevents cleavage by the endogenous restriction enzyme.

55. C is correct.

RNA polymerase is an enzyme that produces primary transcript RNA in transcription.

Molecule C has a phosphodiester bond from the 3' of the base to the 5' downstream base and a triphosphate at the 5' end of the molecule.

A: represents DNA because of the absence of hydroxyl at the 2' position of the sugar.

B: contains a monophosphate at the 5' position.

D: shows a phosphodiester bond at the 2' position (not the 3').

E: shows a phosphodiester bond between two 5' ends.

First step, the two strands of the DNA double helix are physically separated at a high temperature in DNA melting.

Second step, the temperature is lowered, and two DNA strands become templates for DNA polymerase to selectively amplify the target DNA.

Third step, reaction mechanism uses RNA polymerase to synthesize a complementary strand to the template.

Selectivity of PCR results from using *primers complementary* to the DNA region targeted for amplification under specific *thermal cycling conditions*.

The process continues for 30 to 40 cycles, doubling the amount of DNA each cycle.

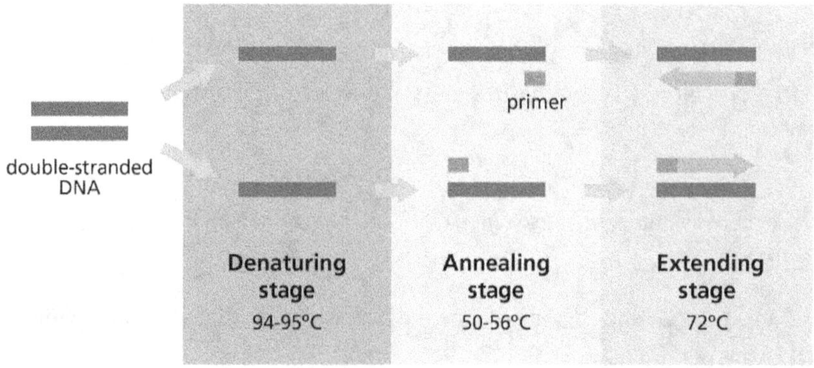

PCR amplification uses three steps for denaturing, annealing, and replicating DNA

57. A is correct.

DNA is double-stranded with A, C, G and T, while RNA is single-stranded with U replacing the T of DNA.

DNA uses the sugar of deoxyribose (i.e., the absence of ~OH group at the 2' in RNA), while RNA uses ribose (i.e., the presence of ~OH group at the 2').

58. B is correct.

Ligase is an enzyme used by the cell during DNA replication (and in other biochemical processes) that catalyzes the joining of two large molecules (e.g., DNA nucleotides) by forming a new chemical bond.

Newly formed bond is *via* a condensation reaction (joining) and usually involves dehydration with the loss of H_2O when the molecules (e.g., DNA, amino acids) are linked.

59. E is correct.

In DNA, thymine (T) base pairs with adenine (A) with two hydrogen bonds, while cytosine (C) base pairs with guanine (G) with three hydrogen bonds.

More energy is required to break three hydrogen bonds than two hydrogen bonds. A DNA sequence with increased C-G pairs has a higher melting point and requires more energy to denature (i.e., separate).

When bonded to its complementary strand, the DNA strand with GCCAGTCG has two T-A and six C-G pairs:

$$(2 \text{ pairs} \times 2 \text{ H bonds} = 4) + (6 \text{ pairs} \times 3 \text{ bonds} = 18) = 22 \text{ H bonds}$$

Thus, this DNA strand has the most hydrogen bonds and the highest melting point.

A: five T-A pairs and three C-G pairs:

$$(5 \text{ pairs} \times 2 \text{ H bonds} = 10) + (3 \text{ pairs} \times 3 \text{ bonds} = 9) = 19 \text{ H bonds}$$

C: four A-T pairs and four C-G pairs:

$$(4 \text{ pairs} \times 2 \text{ H bonds} = 8) + (4 \text{ pairs} \times 3 \text{ bonds} = 12) = 20 \text{ H bonds}$$

D: four A-T pairs and four C-G pairs:

$$(4 \text{ pairs} \times 2 \text{ H bonds} = 8) + (4 \text{ pairs} \times 3 \text{ bonds} = 12) = 20 \text{ H bonds}$$

60. C is correct.

Reverse genetics analyzes the function of a gene by observing the phenotypic effects of specific gene sequences obtained by DNA sequencing.

Reverse genetics seeks to find *which phenotypes* arise due to genetic sequences.

It proceeds in the opposite direction of classical genetics, which investigates the genetic basis of a phenotype.

61. D is correct.

Protein synthesis requires biochemical energy from ATP or GTP.

Two high-energy phosphate bonds (from ATP) provide the energy required to form one aminoacyl-tRNA involving the attachment of each amino acid to its tRNA.

Initiation complex formation requires energy from one GTP.

Initiation requires one high-energy phosphate bond (ATP) is required.

Elongation with the delivery of each new tRNA to the A site requires one GTP.

Translocation of the peptidyl-tRNA requires one GTP.

Termination does not require the hydrolysis of a high-energy phosphate bond (e.g., ATP, GTP).

For each amino acid added to the polypeptide chain, two high-energy phosphate bonds are used for "charging" the tRNA with the correct amino acid (2 GTP × 50 amino acids = 100).

For chain formation, one high-energy phosphate bond is required to carry the amino acid to the ribosome and another to translocate the ribosome (2 GTP × 49 peptide bonds = 98).

Total: 1 + 100 + 98 = 199

62. E is correct.

mRNA in *E. coli* cells is composed primarily of phosphodiester linkages between ribonucleotides.

63. B is correct.

DNA repair is a collection of processes by which a cell identifies and corrects damage to the DNA molecules encoding its genome.

Before replication of DNA, the cell methylates the parental strand for reference during any potential errors introduced during replication.

Normal metabolic activities and environmental factors such as UV light and radiation can cause DNA damage, resulting in many individual lesions per cell per day.

Many lesions cause structural damage to the DNA molecule and can alter the cell's ability to transcribe genes for the survival of its daughter cells after mitosis.

DNA repair process is constantly active as it responds to damage in the DNA structure, and methylation references the original strand when mismatches are detected during replication.

64. A is correct.

Methionine is the *start codon* of mRNA and is the first amino acid in eukaryotic proteins.

Mature protein may excise a portion of the original polypeptide, so methionine is not always the first amino acid in the mature protein after modification in the Golgi.

65. D is correct.

DNA in *E. coli* is composed of four bases (A, T, C, G), phosphodiester linkages connecting deoxyribonucleotide molecules;

two strands base pair in an anti-parallel orientation, and phosphodiester linkages utilizing the 3'-OH.

66. B is correct.

3'–CAGUCGUACUUU–5' anticodon of the tRNA (known from question stem)

5'–GUCAGCAUGAAA–3' codon of the mRNA

3'–CAGTCGTACTTT–5' DNA

There is a polarity when nucleic acids base pair, whereby the 3' end of the tRNA anticodon corresponds to the 5' end of the mRNA codon. C pairs with G, and U pairs with A (base pairing is U to A in RNA).

mRNA sequence is 5'–GUCAGCAUGAAA–3'.

3' end of the RNA hybridizes with the 5' end of DNA.

For complementary DNA, A pairs with T (not U in RNA), and the *polarity* of the strands is *antiparallel.*

An approach: since the 3' end of tRNA and the 3' end of DNA hybridize to the 5' of mRNA, the DNA sequence is the same orientation and similar to tRNA (i.e., replace T in DNA with U in tRNA).

By convention, nucleic acids are written with the 5' end on the left (top left in a double-stranded molecule) and read in the 5' to 3' direction.

67. A is correct.

Promoter is a region of 100–1000 base pairs long on the DNA that initiates transcription of a particular gene.

Promoters are near the genes they transcribe, on the same strand, and upstream on the DNA (towards the 3' region of the antisense strand – template and non-coding strand).

Promoters contain specific DNA sequences and response elements that provide a secure initial binding site for RNA polymerase and transcription factors proteins.

Transcription factors have specific activator (or repressor) sequences of nucleotides that attach to specific promoters and regulate gene expressions.

In bacteria, the promoter is recognized by RNA polymerase and an associated sigma factor (i.e., a protein needed only for initiation of RNA synthesis), which are often brought to the promoter DNA by an activator protein's binding to its own nearby DNA binding site.

In eukaryotes, the process is complicated, with many factors needed to bind RNA polymerase II to a promoter.

68. A is correct.

Isoleucine-glycine is composed of two amino acids and therefore is a dipeptide.

69. E is correct.

Correct order of events in delivering a protein to its cellular destination:

signal sequence binds to a docking protein → membrane channel forms → chaperonins unfold

70. D is correct.

RNA polymerase synthesizes the new strand in the anti-parallel orientation with new nucleotides adding to the growing chain in the 5' to 3' direction.

71. B is correct.

Adenosine (A) bonds with thymine (T) by two hydrogen bonds.

Guanine (G) bonds with cytosine (C) by three hydrogen bonds.

72. B is correct.

Lagging strand is composed of Okazaki fragments.

Nucleotides (e.g., DNA and RNA) extend from the 3'-OH group.

Leading (i.e., continuous) *and* lagging (i.e., discontinuous) strands use the 3'-OH ends as a *nucleophile* during condensation (i.e., via dehydration) reaction for chain elongation.

A: 3' end of the *template strand.*

C: 5' end of a *lagging strand.*

D: 5' end of the *template strand*

73. C is correct.

Ribosomal subunits, radio-labeled with *heavy* carbon and *heavy* nitrogen, were placed in a test tube during bacterial protein synthesis.

Small and large ribosomal subunits assemble to form a complete ribosome during translation.

After translation ceases, the complete ribosome dissociates into individual small and large subunits.

Since the sample used in centrifugation was taken after translation, the individual ribosomal subunits (not the assembled ribosomes) were present.

Centrifugation separates cellular components based on density, and since the subunits are varied sizes, two different bands are expected in the centrifuge tube.

Understanding the size of the two ribosomal subunits in bacteria is required; bacteria have two subunits of 30S and 50S, which assemble to form a 70S complex, and eukaryotes have 40S and 60s ribosomes, which assemble to form an 80S complex.

74. E is correct.

Translation is the protein production process whereby one amino acid adds to the end of a protein.

A ribosome performs this mechanism.

Sequence of nucleotides in the template mRNA chain determines the sequence of amino acids in the generated amino acid chain.

Adding an amino acid occurs at the C-terminus of the peptide, and translation is amino-to-carboxyl directed.

75. D is correct.

Codon for histidine is 5'-CAU-3'. The anticodon in tRNA that brings histidine to the ribosome is 5'-AUG-3'.

76. B is correct.

The existing polypeptide chain is transferred to the P site during translation elongation as the ribosome moves in the 3' direction.

77. A is correct.

In polymerase chain reactions (PCR amplification), a primer hybridizes to the end of a DNA fragment.

A primer is the initiation site for DNA polymerase to bind and replicate the entire strand.

DNA replicates $5' \rightarrow 3'$.

The primer must be the complement of the $3'$ end because DNA polymerase reads the template strand $3' \rightarrow 5'$.

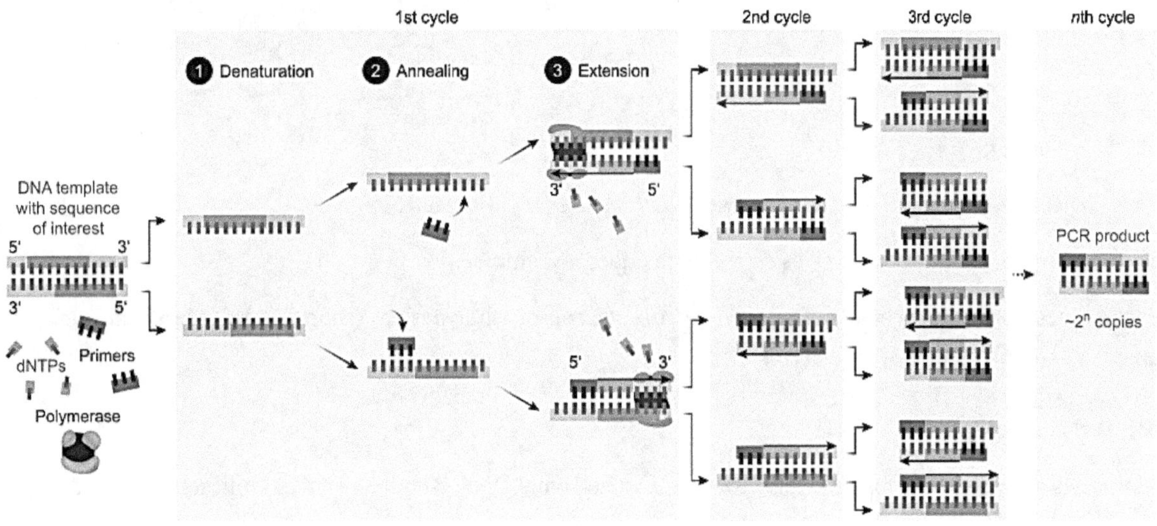

Polymerase chain with exponential product amplification during each cycle. First step heats the DNA to separate the two strands of the double helix. Step 2 cools the sample to allow annealing by complementary primers. Step 3 is chain elongation for the synthesis of DNA by RNA polymerase extending from the primers.

78. C is correct.

Genetic code is *not* ambiguous (i.e., each codon specifies one amino acid).

79. C is correct.

R-groups (i.e., amino acid side chains) portion of the polypeptide chain is responsible for establishing and maintaining the force to stabilize the secondary structure.

80. B is correct.

A ribosome made a tripeptide, MET-ARG-SER, attached to tRNA in the P site.

From the genetic code, the CGU codon is in the E site of the ribosome.

81. E is correct.

A ribosome made a tripeptide, MET-ARG-SER, attached to tRNA in the P site.

Using the genetic code, it cannot be determined which codon is in the A site of the ribosome.

First base (5' end)	Second base U	Second base C	Second base A	Second base G	Third base (3' end)
U	UUU Phe	UCU	UAU Tyr	UGU Cys	U
	UUC	UCC Ser	UAC	UGC	C
	UUA Leu	UCA	UAA Stop	UGA Stop	A
	UUG	UCG	UAG Stop	UGG Trp	G
C	CUU	CCU	CAU His	CGU	U
	CUC Leu	CCC Pro	CAC	CGC Arg	C
	CUA	CCA	CAA Gln	CGA	A
	CUG	CCG	CAG	CGG	G
A	AUU	ACU	AAU Asn	AGU Ser	U
	AUC Ile	ACC Thr	AAC	AGC	C
	AUA	ACA	AAA Lys	AGA Arg	A
	AUG Met start	ACG	AAG	AGG	G
G	GUU	GCU	GAU Asp	GGU	U
	GUC Val	GCC Ala	GAC	GGC Gly	C
	GUA	GCA	GAA Glu	GGA	A
	GUG	GCG	GAG	GGG	G

82. E is correct.

Three *pyrimidines* are *cytosine* and *thymine* (*uracil* replaces thymine in RNA).

The word pyrimidine contains "y" as do the nucleotides, pyrimidines.

Pyrimidines (longer word than purine) are *one-ring structures*, while purines (shorter word) are adenine and guanine, larger structures with *two rings*.

83. D is correct.

Plasmid is a small DNA molecule separate from chromosomal DNA within a cell and replicates independently.

Commonly small, circular, double-stranded DNA molecules in bacteria, plasmids are sometimes in archaea and eukaryotic organisms.

In nature, plasmids carry genes that may benefit the organism's survival (e.g., antibiotic resistance in bacteria) and can be transmitted among bacteria, even of another species.

Artificial plasmids are commonly used as vectors in molecular cloning to drive the replication of recombinant DNA sequences within host organisms.

84. A is correct.

Plasmids are small circular double-stranded DNA in bacteria carrying *extrachromosomal genetic information.*

Plasmids (with an origin of replication) are replicated by bacterial proteins and inherited by progeny.

Proteins encoded by genes on plasmids often provide resistance to antibiotics by degrading the antibiotic.

Through recombinant DNA technology, plasmids are engineered to carry other genes *not typically in bacteria.*

Plasmids are introduced by *transforming* bacterial cells (i.e., through the cell wall and plasma membrane.

After transformation (i.e., uptake of DNA), exposure to the specific antibiotic (e.g., penicillin) allows the selection of bacteria transformed with the plasmid by clone selection (i.e., antibody-resistant clones).

85. A is correct.

Gene therapy focuses on genetically modifying cells to produce a therapeutic effect (or the treatment of disease) by repairing or reconstructing defective genetic material.

Notes for active learning

Notes for active learning

Notes for active learning

Genetics and Inheritance Patterns – Detailed Explanations

1. A is correct.

Model organism for *genetic studies* (e.g., pea plants, Drosophila, zebrafish) has common features. The organism must be bred in *large numbers*, and *ease of cultivation* favors viable offspring that *transmit genetic information* between generations.

Genetic studies rely upon statistics that favor *large sample numbers*. Therefore, a short generation time (i.e., the span between birth and fecundity) is preferred.

There should be *discreet phenotypic differences* among alleles (alternative forms of the gene).

For example, among the seven traits Mendel (1822-1884) observed, he inventoried tall *vs*. short plants, round *vs*. smooth seeds, and green *vs*. yellow seeds.

Mendel controlled the crosses by manually transferring pollen from the anther of a mature pea plant of one variety to the stigma of a separate mature pea plant of the second variety.

Organisms should have a *well-characterized genome* (gene identity and function have been studied).

Increasing *numbers of chromosomes* increase the genes that can influence the observable phenotypic outcomes.

Fewer chromosomes facilitate statistical analysis and causation when the genome is manipulated.

2. C is correct.

Sex-linked genetic disease, hemophilia, causes excessive bleeding because the blood does not clot. Tom, Mary, and their four daughters do not exhibit symptoms of hemophilia.

However, their son exhibits symptoms of hemophilia because Mary is heterozygous.

3. C is correct.

Eye color is sex-linked in *Drosophila*.

Determine the phenotype of parents.

Red-eyed flies with red-eyed and sepia-eyed parents must be heterozygous because a sepia-eyed parent only contributes recessive sepia alleles.

When a heterozygous (Rr) red-eyed fly is crossed with a homozygous recessive (rr) sepia-eyed fly, ½ of the offspring are red-eyed (Rr) because of the dominant (red) allele from the heterozygous fly.

Punnett Square:

Red eyed parent

Sepia eyed Parent	R	r
r	Rr (red)	rr (sepia)
r	Rr (red)	rr (sepia)

continued...

Since the question does not assign gender to the sepia and red-eyed parents, the Punnett squares for two combinations for sex-linked traits are:

		Red eyed female (♀)	
		R	r
Sepia eyed male (♂)	r	Rr (red)	rr (sepia)
	y	Ry (red)	ry (sepia)

		Red eyed male (♂)	
		R	y
Sepia eyed female (♀)	r	Rr (red)	ry (sepia)
	r	Rr (red)	ry (sepia)

4. E is correct.

Color blindness pertains to cone photoreceptors in retinas, as the cones can detect the color frequencies of light.

About 8 percent of males, but 0.5 percent of females, are colorblind, whether it is one color, a color combination, or another mutation.

Males are at a greater risk of inheriting an X-linked mutation because males have one X chromosome (XY) while females have two (XX).

Men lack a second X chromosome to compensate for the X chromosome that carries the gene mutation.

If a woman inherits a typical X chromosome in addition to the one that carries the mutation, she does not display the mutation.

5. C is correct.

Microscopy is the laboratory technique of magnifying objects that cannot be seen with the unaided eye.

Gregor Mendel (1822-1884) performed crossbreeding experiments with pea plants to study inheritance patterns and introduced the terms *dominant* and *recessive* for phenotypic traits (or *alleles*).

6. E is correct.

AAbbCc produces 2 gametes:

AbC and Abc = 1/2

AaBbCc produces 8 gametes:

ABC, ABc, AbC, Abc, aBC, aBc, abC, abc = 2/8 possible = 1/4

From probability: $1/2 \times 1/4 = 1/8$

	ABC	ABc	AbC	Abc	aBC	aBc	abC	abc
AbC	X	X	X	X	X	**Yes**	X	X
Abc	X	X	X	X	**Yes**	X	X	X

7. B is correct.

Afflicted children are aa = 1 of 4 possibilities = ¼ or 25%.

	A	a
A	AA	Aa
a	Aa	aa

8. C is correct.

Tumor suppressor genes protect a cell from aberrant cell cycles.

When this gene's function is lost or reduced due to mutation, the cell can progress to cancer (usually combined with other genetic mutations).

Loss of tumor suppressor genes may be more critical than proto-oncogene/oncogene activation to form many kinds of human cancer cells.

Both alleles of the tumor suppressor gene encoding a particular protein must be affected before an effect is manifested.

If one allele is damaged, the second can produce the correct protein.

9. D is correct.

Color blindness is a *sex-linked trait* because the gene is on the X chromosome.

Mother is a carrier (not afflicted with condition) and is heterozygous for the recessive allele (color blindness).

Father has the allele on his X chromosome (Y chromosome lacks the gene).

Genotype and *phenotype* of an XY son depend entirely on the mother (afflicted vs. carrier) since the afflicted father transmits the gene on his X.

Mother is heterozygous; a son has a 50% probability of receiving the color blindness allele from his mother.

10. B is correct.

If two strains of true-breeding plants with different alleles are crossed, their progeny is the F_1 generation.

11. C is correct.

AaBbCcDdEe × AaBbCcDdEe

From probability:

$\frac{1}{2} \times \frac{1}{2} \times \frac{1}{2} \times \frac{1}{2} \times \frac{1}{2} = 1/32$

12. A is correct.

Autosomal recessive inheritance is the product of mating two carriers (i.e., heterozygous parents).

In mating two heterozygotes for an autosomal recessive gene, there is a:

25% (1/4) probability of a homozygous *unaffected* child

25% (1/4) probability of a homozygous *affected* child

50% (1/2) probability of a heterozygous (*carrier*) child

75% of children are phenotypically normal (25% AA and 50% Aa).

Of all children, 50% are phenotypically normal but carry the mutant gene (Aa).

13. D is correct.

If the dominant allele frequency is three times that of the recessive allele,

$$p = 3q$$

Hardy-Weinberg equilibrium:

$$p + q = 1$$

so

$$3q + q = 1$$

Solving for q

$$4q = 1$$

$$q = 0.25$$

$$p = 0.75$$

Allele frequency:

$$p^2 + 2pq + q^2 = 1$$

Heterozygote allele = 2pq

Substituting for p and q,

$$2(0.75)\cdot(0.25) = 0.375 \text{ or } 37.5\%$$

14. E is correct.

Loss of heterozygosity is a chromosomal event resulting from losing the gene and surrounding region.

Diploid cells (e.g., human somatic cells) contain *two copies* of the genome, one from each parent.

Each copy contains approximately 3 billion bases, and for most positions in the genome, the base is consistent between individuals. However, a small percentage may contain different bases.

These positions are *single nucleotide polymorphisms* (or SNP). The region is heterozygous when the genomic copies from each parent have different bases for these regions.

Most chromosomes within somatic cells are paired, allowing SNP locations to be potentially heterozygous.

One parental copy of a region can be lost, resulting in the region with just one copy.

If the copy lost contained the dominant allele, the remaining recessive allele would appear in a phenotype.

15. A is correct.

Maternal inheritance involves all progeny exhibiting the phenotype of the *female* parent.

B: not maternal inheritance because the progeny exhibits the phenotype of the *male* parent.

C and D: Mendelian 1:1 segregation and not maternal inheritance.

Maternal inheritance is uniparental when all progeny have the genotype and phenotype of the female parent.

16. E is correct.

Gene is a fundamental physical, functional unit of heredity transferred from a parent to offspring and determines some offspring characteristics.

Genes are DNA sequences encoding proteins.

Alleles are forms of the same gene with slight differences in their sequence of DNA bases.

Genome is the *complete set of genetic information* for an organism.

Genome has the genetic information needed for an organism and allows it to develop, grow and reproduce.

17. B is correct.

GC base pairs converted to AT base pairs in the promoter will likely lose gene function completely.

18. A is correct.

Mutations affect proteins but not lipids or carbohydrates.

In proteins, the effects on the protein are no change (i.e., silent mutation), abnormal protein production, loss of protein (enzyme) function, or gain of protein (enzyme) function.

Loss of function of a gene product may result from mutations encoding a regulatory element or the loss of critical amino acid sequences.

continued...

Gain-of-function mutations are changes in the amino acids resulting in enhancement of the protein function.

There may be an increase in the level of protein expression (affecting the operator region of the gene) or an increase in each protein molecule's ability to perform its function (change in the shape of the protein).

19. A is correct.

The desired phenotype is green smooth peas, and green and smooth are dominant phenotypes.

Therefore, the genotypes selected for the cross must avoid the two recessive alleles (g and s).

For GgSs × GGSS, one parent (GGSS) is a double dominant, and therefore all offspring have the dominant phenotype (G and S) regardless of the other parent's genotype.

B: Gg × gg yields 1/2 yellow (g) phenotype offspring

C: ss × Ss yields 1/2 wrinkled (s) phenotype offspring

D: Gg × Gg yields 1/4 yellow (g) phenotype offspring

20. C is correct.

Retinoblastoma (Rb) is a rapidly developing cancer in immature cells of the retina, the light-detecting tissue of the eye. It is a common malignant tumor of the eyes in children. A single allele is inherited (i.e., dominant) for the phenotype.

21. E is correct.

Phenotype of the first child does not influence the probability of the second child.

For example, the probability of getting a tail on the first toss of a coin does not influence the probability of getting a head on the second toss.

Punnett square determines possible gametes and their combinations.

If ½ of the woman's gametes carry the trait and ½ of the father's gametes carry the trait, the probability of a child receiving the allele from each parent is ½ × ½ = ¼.

22. C is correct.

In Mendel's experiment, the cross of spherical-seeded and wrinkled-seeded pea plants inherited alleles (or *gene variants*) from each parent.

However, only spherical-seeded plants resulted from the cross.

Wrinkled-seed gene is a recessive allele compared to the spherical-seed gene (i.e., the dominant gene).

Dominant allele is a genetic variant expressed more strongly than other variants (or alleles) of the gene (i.e., recessive) for many reasons.

23. B is correct.

Notation 2N indicates that a given cell line is diploid, two homologous versions of each chromosome.

Human somatic cells are diploid with 23 different chromosome pairs (N = 23) for 46 chromosomes (2N = 46).

Gamete cells (eggs and sperm) are haploid (i.e., 1N).

Mitosis is the mode of cell division used by somatic cells, resulting in two diploid daughter cells genetically identical to the diploid parent cell.

24. D is correct.

Degree of genetic linkage measures the physical distance of two genes on the same chromosome.

Probability of crossover and corresponding exchange between gene loci (location on the chromosome) is generally proportional to the distance between the loci.

Genes far apart on a chromosome are more likely to be separated during crossover than genes physically close.

Thus, *frequency of genetic recombination* between two genes is *related to their distance.*

Recombination frequencies are used to construct genetic maps.

One map unit (Morgan units) is defined as a 1 percent recombinant frequency.

Recombination frequencies are roughly *additive* and are a good approximation for small percentages.

Largest percentage of recombinants cannot exceed 50%, resulting when the two genes are at the opposite ends of the same chromosome.

Crossover events result in an exchange of genes.

However, an odd number of crossover events (a 50% probability between an even and an odd number of crossover events) results in a recombinant product.

25. A is correct.

Epigenetic inheritance results from changes in gene activity, *not* caused by changes in the DNA sequence.

It studies stable, long-term alterations in the transcriptional potential that are not necessarily heritable.

Unlike simple genetics based on changes to the DNA sequence (genotype), the changes in gene expression or cellular phenotype of epigenetics have other causes.

For example, *cellular differentiation* is an epigenetic change in eukaryotes cells.

During *morphogenesis*, totipotent (i.e., all potent) stem cells become pluripotent (i.e., highly potent, but with limited determinate potential) cells of the embryo, which become fully differentiated cells.

Gene expression is when a single fertilized egg cell (zygote) divides, and the resulting daughter cells change into different cell types (e.g., neurons, muscle cells, epithelium, endothelium of blood vessels) by activating some genes while inhibiting the expression of others.

26. C is correct.

Female children receive one X from their mother and one X from their father.

X from the father must carry the color blindness allele because the father is colorblind.

X from the mother has a wild-type and a color blindness allele because she is heterozygous recessive.

50% of female children are homozygous colorblind, and 50% are heterozygous carriers.

27. D is correct.

True breeding means that the organism is homozygous (e.g., AA or aa) for the trait.

All progeny are heterozygous Aa (below) and exhibit the dominant phenotype.

	A	A
a	Aa	Aa
a	Aa	Aa

28. C is correct.

	A	A
a	Aa	Aa
a	Aa	Aa

	A	a
A	AA	Aa
a	Aa	aa

F_1: all tall F_2: ¾ are tall, and ¼ is short

29. B is correct.

Frameshift mutation is when 1 or 2 base pairs are added or deleted. A 3 base pair addition or deletion causes an *in-frame* mutation because 3 nucleotides encode each codon.

Frameshift mutation (i.e., addition/deletion of other than multiples of 3 nucleotides) causes the ribosome to read all downstream codons in the wrong frame. They usually result in truncated (i.e., *nonsense* mutation) or non-functional proteins (i.e., *missense* mutation).

An altered base pair (i.e., *point* mutation) is not a frameshift mutation because it substitutes (not adds or deletes) and does not cause the ribosome to read codons out of frame.

Point mutations can result in nonsense (i.e., premature stop codon) or missense (i.e., improperly folded protein) mutations.

Base pair additions/deletions (other than in multiples of 3) cause frameshift mutations.

30. D is correct.

Mendel's *law of independent assortment* states that the probability of a cross resulting in a genotype equals the *product* of individual probabilities.

Crosses by two heterozygous individuals for the three genes (A, B, and C) produce homozygous dominant offspring for each trait.

Each parent is heterozygous for A, genotype = Aa.

Ratio of offspring equals 1/4 AA, 1/2 Aa, and 1/4 aa; a typical 1:2:1 ratio for heterozygous crosses.

Parents are heterozygous for genes B and C, the probability of offspring being BB is 1/4, and CC is 1/4.

Probability that offspring are genotype AABBCC = *product of individual probabilities*:

$$1/4 \times 1/4 \times 1/4 = 1/64$$

31. E is correct.

Nonsense mutation is a DNA point mutation resulting in a premature stop codon in the transcribed mRNA and a truncated (i.e., incomplete) protein, usually nonfunctional.

Missense mutation is a point mutation where a nucleotide is changed and substitutes for a different amino acid.

Genetic disorders of sickle cell anemia, thalassemia and Duchenne muscular dystrophy arise from *nonsense mutations*.

32. B is correct.

Recessive trait is expressed when present in both copies (i.e., alleles) or is the single copy of the gene.

Human Y chromosome confers maleness.

Recessive (single copy) alleles on the X are expressed (e.g., hemophilia).

Recessive X-linked allele is expressed in unaffected females with two (homozygous) alleles.

33. B is correct.

Mendelian *Laws of inheritance* explain patterns of disease transmission.

Inheritance patterns of single-gene diseases are *Mendelian* after Augustinian friar Gregor Mendel (1822-1884), who first reported different gene segregation patterns for specific garden peas traits.

Mendel calculated *probabilities* of inheritance for traits in the next generations.

Because of mutations or polymorphisms, most genes have more versions (i.e., alleles).

Individuals carry normal, mutant, or rare alleles, depending on mutation/polymorphism and allele frequency within a population.

Single-gene diseases are usually inherited depending on gene location and whether one or two regular copies of the gene are needed for the disease to manifest (i.e., affected individuals).

continued...

Expression of a mutated allele is *dominant*, *co-dominant*, or *recessive*.

Five basic *patterns of inheritance for single-gene diseases*:

Autosomal dominant:

each affected person has an affected parent

manifests in each generation

Autosomal recessive:

parents of an affected person are carriers (i.e., unaffected)

typically, NOT seen in each generation

X-linked dominant:

females affected more frequently

can affect males and females in the same generation

X-linked recessive:

males affected more frequently

often affects males in each generation

Mitochondrial:

males and females can be affected but passed by females

can appear in each generation

An accurate family history is essential to determine inheritance pattern when a family is affected by a disease.

34. A is correct.

Two traits are *unlinked* when inherited on separate chromosomes or because the genes are apart greater than 50 centimorgans.

At large distances, double-crossing over occurs, and the genes appear unlinked.

35. D is correct.

Let T = tall and t = short; B = brown eyes and b = blue eyes.

Father is *homozygous tall* and *blue-eyed*; his genotype is TTbb.

Mother is *heterozygous tall* and *heterozygous brown-eyed*; her genotype is TtBb.

Determine the probability that parents produce a tall child with blue eyes (T_bb).

The genes for height and eye color are unlinked.

Father (TTbb) contributes T and b alleles, so his gametes have T and b alleles.

Mother (TtBb) contributes T or t and B or b, so her gametes are (in equal amounts): TB, tB, Tb, or tb.

continued...

Genotypes of the offspring:

TTBb, TTbb, TtBb, Ttbb

Half the offspring are tall and brown-eyed (T_B_), and half are tall and blue-eyed (T_bb).

Therefore, the probability of a tall child with blue eyes is ½.

A faster method is calculating phenotype ratios for height and eye color separately and then combining them.

Mating TT × Tt = 100% tall

Mating Bb × Bb = ½ blue and ½ brown

Multiplying 1 tall × ½ blue = ½ tall blue

36. A is correct.

Recombination is the exchange of genetic information between homologous chromosomes and occurs during prophase I of meiosis.

Crossing over between non-sister homologs in *meiosis* results in a new combination of alleles, for example:

AB / ab can yield Ab / aB

Recombination occurs in eukaryotes during *mitosis* but between *sister chromatids* which are copies (replicated during the S phase) and therefore do *not* lead to novel genotypes.

Recombination is a *DNA repair mechanism* between homologous chromosomes.

Research supports that recombination is not a random event.

Recombination hotspots include chromosome regions with *high GC content* and particular architecture (e.g., genome size, haploid chromosome number, chromosome size, and chromosome rearrangements).

Enzymes catalyze recombination (e.g., rec A, rec B, rec C, and rec D) by initiating and *facilitating strand invasion* and *strand transfer* during recombination.

High recombination frequency means the genes are *farther apart.*

Each percent frequency of recombination equals one map unit between the genes. So, 2.5% recombination frequency equals genes 2.5 map units apart.

Largest recombination frequency is 50%, as if the genes were on different chromosomes (consistent with Mendel's Law of Independent Assortment).

Recombination frequency would be the same for *cis*- and *trans*-heterozygotes because the distance is the same between the genes regardless of whether they are on the same (*cis*) or homologous (*trans*) chromosomes.

Recombination frequency is *not* a completely random event.

Specific regions within the chromosome have differences in the propensity to undergo recombination; the presence of *hotspots* and architectural features (e,g, histones) increase or decrease recombination frequencies.

Recombination frequency increases (not decreases) with *distance.*

Genes have different distances; with different distances, the recombination frequency changes.

37. C is correct.

There are two possible alleles for each of the three genes.

If the genes assort independently (not linked):

$2^3 = 8$ combinations exist

38. A is correct.

Each affected person has an affected parent in the *autosomal dominant* inheritance pattern.

Autosomal dominant inheritance is a way a genetic trait can be passed from parent to child.

One parent's copy of a mutated (changed) gene can cause genetic conditions.

A child who has a parent with the mutated gene has a 50% chance of inheriting that mutated gene.

Men and women are *equally likely* to have these mutations.

Sons and daughters are *equally likely* to inherit them.

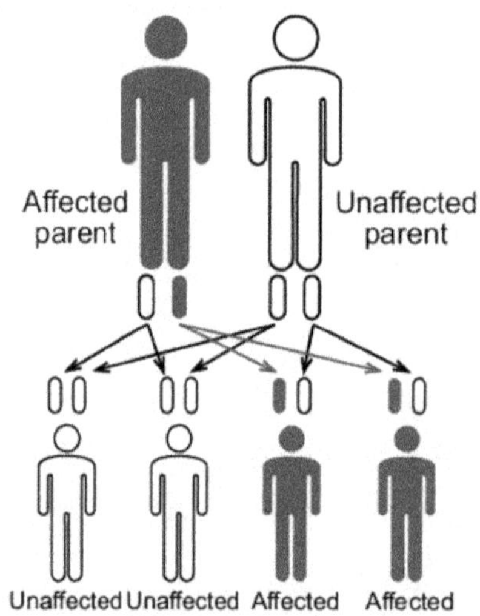

Autosomal dominant inheritance pattern with presence in each generation

39. C is correct.

Parents are Aa × Aa (carriers but not afflicted with the disease).

Progeny could be AA, Aa, Aa, or aa.

From the Punnett square, eliminate aa because this is the disease state.

The question asks the probability that she is heterozygous (Aa) but not homozygous (AA) = 2/3.

40. E is correct.

Anticipation is associated with an earlier onset of symptoms and increased disease severity in each generation.

Anticipation is observed in autosomal dominant diseases by *trinucleotide repeat expansions* (e.g., Huntington's disease, myotonic dystrophy) because increased-length triple repeats are unstable during cell division.

A: *codominance* is when a gene has more than one dominant allele. In the ABO blood group system, the I^A and I^B alleles are codominant.

Heterozygous individual for *two codominant alleles* expresses the phenotypes associated with both alleles.

Heterozygous individuals for the I^A and I^B alleles express the AB blood group phenotype, A- and B-type antigens are on the surface of red blood cells.

Codominance occurs at the locus for the beta-globin component of hemoglobin.

Three molecular phenotypes of Hb^A/Hb^A, Hb^A/Hb^S, and Hb^S/Hb^S are detectable by protein electrophoresis.

B: *penetrance* is the proportion of individuals carrying the variant of a gene (allele or genotype) and expressing the trait (phenotype).

In medical genetics, the penetrance of a disease-causing mutation is the proportion of individuals with the mutation who exhibit clinical symptoms.

For example, if a mutation in the gene responsible for an autosomal dominant disorder has 95% penetrance, 95% of those with the mutation develop the disease, while 5% do not.

D: *gain of function mutations* change the gene product to gain a new and abnormal function.

These mutations usually have dominant phenotypes and are expressed with a single allele.

41. B is correct.

Hardy-Weinberg equation:

$$p^2 + 2pq + q^2 = 1$$

where p equals gene frequency of dominant allele, and q equals gene frequency of recessive allele

Hence, in the population:

p^2 is the frequency of *homozygous dominants*

2pq is the frequency of *heterozygotes*

q^2 is the frequency of *homozygous recessives*

For a trait with two alleles, p + q must equal 1 since the combined frequencies of the alleles = 100%.

If the frequency of the recessive allele for a trait is 0.6,

$$q = 0.6$$

continued...

Since

$$p + q = 1$$

$$p = 0.4$$

To calculate the frequency of individuals expressing the *dominant phenotype* (not the dominant genotype), determine the number of individuals *homozygous for the dominant trait* (p^2) and add the *number of heterozygotes* (2pq) exhibiting the dominant phenotype:

$$p^2 = (0.4) \times (0.4)$$

$$p^2 = 0.16$$

$$2pq = 2 \times (0.6) \times (0.4)$$

$$2pq = 0.48$$

So,

$$p^2 + 2pq = 0.16 + 0.48$$

$$p^2 + 2pq = 0.64$$

A: 0.48 = frequency of *heterozygous* individuals.

C: 0.16 = frequency of *homozygous dominant* individuals.

D: 0.36 = frequency of *homozygous recessive* individuals.

42. C is correct.

Maximum recombination frequency between two genes is 50%.

43. B is correct.

For a recessive trait to appear in a phenotype of an offspring (e.g., long hair), offspring inherit a recessive allele (Mendel called it *traits*) from each parent (i.e., two copies of the recessive gene).

Short-haired parents carry one copy (i.e., heterozygous) of the recessive long-haired allele (or *gene*).

Combined with the second copy from the other long-haired parent, it produced the long-haired offspring.

44. D is correct.

Meiosis in males produces four haploid (1N) unique cells genetically different from the parental cell.

45. D is correct.

Point mutations occur when a single nucleotide base (A, C, G, T) is substituted by another.

Silent mutation is a point mutation that

1) occurs in a noncoding region or

2) does not change the amino acid sequence due to the degeneracy of the genetic code.

Frameshift mutation is the insertion or deletion of some nucleotides. These mutations severely affect the coded protein since nucleotides are read as triplets.

Addition or loss of nucleotides (except in multiples of three) changes the reading frame of the mRNA and often gives rise to premature polypeptide termination (i.e., nonsense mutation).

Missense mutation results from the insertion of a single nucleotide that changes the amino acid sequence of the specified polypeptide.

46. E is correct.

Two reciprocal crossing-over events appear in progeny at an approximate ratio of 1:1.

47. C is correct.

EEBB × eebb produces offspring of single genotype EeBb, as determined by the Punnett square for the cross between a homozygous dominant by a homozygous recessive.

58. D is correct.

Mutations may cause *premature translation termination* (i.e., *nonsense* mutation) and nonfunctional protein.

49. E is correct.

	A	a
A	**AA**	**Aa**
a	**Aa**	aa

Afflicted children are AA, Aa or Aa (not aa) = 3 of 4 possibilities = ¾ or 75%.

50. D is correct.

Recombinant frequencies of linked genes map the relative locations of genes on a single chromosome.

Recombinant frequencies are determined by crossing individuals that differ in alleles for the genes in question and determining the genotypes of their progeny.

Recombinant frequencies equal frequencies of nonparent genotypes since these genotypes arise by crossover.

Mapping is based on the *probability* of a crossover between two points; the probability of crossover *increases* as the distance between the genes *increases*.

continued...

Farther away, genes have a *greater* recombinant frequency.

Probability that two genes are inherited together (i.e., linked) *decreases* as the distance between them on a chromosome *increases*.

One map unit = 1% recombination frequency, and recombinant frequencies are (roughly) additive.

However, if the genes are far apart, the recombination frequency reaches a *maximum of 50%,* at which point the genes are considered to be sorted independently.

There are four genes (D, E, F, and G), and the recombinant frequencies between each pair are given.

To construct the map, *start* with the allele pair with the *highest recombinant frequency*: between G and E (23%), which means that G and E are 23 map units apart and on the two ends.

Determine the intervening genes by finding the genes closest to the two endpoints.

G and D are 8 map units apart, closest to G. Thus, D must be next to G.

Genes on this chromosome must be G, D, F, and E by elimination.

EFDG is equally correct if the map started from the opposite direction, but this is not an answer choice.

To verify, D and E are 15 map units apart because the distance from G to D, which is 8, plus the distance from D to E, which is 15, is the distance from G to E, which equals 23.

G and F are 15 map units apart, while F and E are 8 units apart.

The numbers add, whereby the distance from G to E equals G to D + the distance from D to E.

The observed numbers may be off by one or two map units (not a mistake) because map distances are roughly additive (i.e., based on rounding for the probabilities).

Questions **51** through **57** are based on the following:

The pedigree illustrated by the schematic shows the inheritance of albinism, a homozygous recessive condition manifested in a total lack of pigment. Specify the genotypes using *A* and *a* to indicate dominant and recessive alleles.

Note: solids are albino individuals.

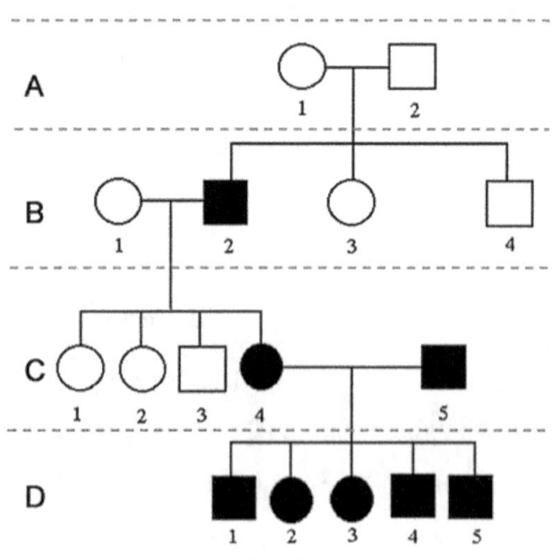

51. C is correct.

Individual A-1 in the pedigree is *Aa.*

52. C is correct.

Individual A-2 in the pedigree is *Aa.*

53. C is correct.

Individual B-1 in the pedigree is *Aa*.

54. B is correct.

Individual B-2 in the pedigree is *aa*.

55. C is correct.

Individual C-3 in the pedigree is *Aa*.

56. B is correct.

Individual C-4 in the pedigree is *aa*.

57. B is correct.

Individual D-4 in the pedigree is *aa*.

58. E is correct.

BBss × bbss produces gametes Bbss.

Then crossed:

Bbss × Bbss and produce the four gametes:

BBss, Bbss, bBss and bbss

Phenotypically,

¾ (BBss, Bbss, bBss) are black and spotted

¼ (bbss) is red and spotted

59. A is correct.

X-linked recessive inheritance is a mode of inheritance whereby a mutation in a gene on the X chromosome causes the phenotype to be expressed

1) *hemizygous males* with a single allele of the mutation because they have one X chromosome and

2) *homozygous females* who are homozygous for the mutation (i.e., a copy of the gene mutation on each of two X chromosomes).

X-linked inheritance indicates that the gene is on the X chromosome.

Females have two X, while males have one X and one Y.

Carrier females have one allele and do not usually express the phenotype.

continued...

X-linked gene mutations are more common in males (i.e., a single allele of the X chromosome) than in females (i.e., two X chromosomes).

Gene appears to *alternate generations* because heterozygous phenotypically normal females transmit it.

Affected males transmit the gene to their daughters, carriers for the trait.

There is no father-to-son transmission because the sons inherit the father's Y chromosome.

Males carrying the mutant gene show the trait.

60. C is correct.

X-inactivation (i.e., lyonization) is when one of two X chromosomes in female mammals is inactivated.

Inactive X chromosome is silenced by transcriptional inactivity within heterochromatin (i.e., condensed DNA).

Females have two X chromosomes, and X-inactivation prevents them from having twice as many X chromosome gene products as males (XY) for dosage compensation of the X chromosome.

Which X chromosome becomes inactivated is random in humans.

Once an X chromosome is inactivated, it remains inactive throughout life and in descendants.

Calico and tortoiseshell-colored cats are phenotypic examples of *X-inactivation* because the alleles for black and orange fur coloration reside on the X chromosome.

For a patch of fur, inactivation of an X chromosome carries one gene resulting in the fur color of the allele for the active gene.

Calico cats are almost always female because the X chromosome determines the color of the cat, and female cats (like all female mammals) have two X chromosomes.

Male mammals have one X and one Y chromosome.

Y chromosome lacks color genes; there is no probability a male cat could have orange and non-orange.

One prominent exception is when, in rare cases, a male has XXY chromosomes (Klinefelter syndrome).

61. D is correct.

Autosomal dominance is typical Mendelian inheritance.

An individual needs a single copy of the mutant gene to exhibit the disease for *autosomal dominant* traits.

Autosomal dominance usually,

equal numbers of males and females affected

traits do not skip generations

father-to-son transmission

62. E is correct.

Deletion of a portion of the Y chromosome containing the *testis-determining factor* results in some XY individuals being phenotypically female.

63. C is correct.

Inbreeding exposes recessive alleles by increasing *homozygosity* by one allele from each parent.

Closely related individuals are likely to carry alleles of *recessive genes* and have an increased probability of *passing recessive alleles* to their offspring.

64. D is correct.

Polyploidy refers to a numerical change in a complete set of chromosomes.

Most eukaryotic organisms are diploid (2N) with two sets of chromosomes – one set inherited from each parent.

Polyploid cells and organisms contain *more than two paired* (i.e., homologous) *sets of chromosomes*.

If crossing between two species is not possible because of differences in ploidy level, polyploids can be used as a bridge for gene transfer.

Additionally, polyploidy reduces fertility due to meiotic errors, allowing the cultivation of seedless varieties.

Polyploidy is in some organisms and is especially common in plants.

In addition, polyploidy occurs in some tissues of animals that are otherwise diploid (e.g., human muscle tissues).

65. C is correct.

Transgenesis introduces an exogenous gene (or *transgene*) into an organism to exhibit a new characteristic transmitted to its offspring.

Hybrid typically refers to the offspring of different species from interbreeding between two animals (or plants).

66. D is correct.

Plasmid is a small circular DNA molecule physically separate from chromosomal DNA within a cell and can replicate independently.

In nature, plasmids carry genes that may benefit the organism's survival (e.g., *antibiotic resistance in bacteria*) and can be transmitted among bacteria, even of another species.

Artificial plasmids are commonly used as vectors in molecular cloning to confer antibiotic resistance and drive the replication of recombinant DNA sequences within host organisms.

67. B is correct.

Transgenesis introduces an exogenous gene (a transgene) into an organism so that the organism develops a new characteristic and transmits it to offspring.

Hybrid in genetics has several meanings, the most common of which is the offspring resulting from the interbreeding between two animals or plants of different species.

68. B is correct.

DNA profiling (or *genetic fingerprinting*) is a laboratory technique to identify individuals by their DNA profiles.

DNA profiles are deoxyribose nucleotide bases (i.e., DNA used for identification.

Genetic fingerprinting (or DNA *profiling*) is used in paternity testing and criminal investigation and differs from complete genome sequencing (e.g., human genome project).

About 99.9% of human DNA sequences are the same.

DNA profiling distinguishes individuals with a high probability of certainty unless they are monozygotic (i.e., identical twins).

69. E is correct.

X-linked dominance is a mode of inheritance whereby a dominant gene is on the X chromosome.

X-linked dominant is less common than X-linked recessive.

For X-linked dominant inheritance, one allele is sufficient to cause the disorder when inherited from a parent who has the disorder.

X-linked dominant traits do not necessarily affect males more than females (unlike X-linked recessive traits).

An affected father has all affected daughters but no affected sons (unless the mother is also affected).

Notes for active learning

Frank J. Addivinola, Ph.D.

This study guide's lead author and chief editor is Dr. Frank Addivinola. With his outstanding education, laboratory research, and decades of university science teaching, Dr. Addivinola lent his expertise to oversee the development of this series.

During his extensive career, Dr. Addivinola held faculty positions at colleges and universities, including Harvard University, Johns Hopkins University, University of Maryland, and Northeastern University, and taught undergraduate and graduate-level courses in biology, biochemistry, organic chemistry, inorganic chemistry, physics, anatomy and physiology, medical terminology, nutrition, and medical ethics. He received several awards for his research and presentations.

Dr. Frank Addivinola conducted original research in developmental biology as a doctoral candidate and pre-IRTA fellow in Molecular and Cell Biology at the National Institutes of Health (NIH). His dissertation advisor was Nobel laureate Marshall W. Nirenberg, Chief of the Biochemical Genetics Laboratory at the National Heart, Lung, and Blood Institute (NHLBI). Before NIH, Dr. Addivinola researched prostate cancer in the Cell Growth and Regulation Laboratory of Dr. Arthur Pardee at the Dana Farber Cancer Institute of Harvard Medical School.

Dr. Addivinola holds an undergraduate degree in biology from Williams College. He completed his Masters at Harvard University, Masters in Biotechnology at Johns Hopkins University, and five other graduate degrees at the University of Maryland University College, Suffolk University, and Northeastern University.

College Cell & Molecular Biology Subject Review provides comprehensive coverage of cell and molecular biology topics taught in standard college course. The content covers all essential principles and concepts necessary to understand the material and do well on college tests and exams.

Visit our Amazon store

College Organismal Biology:
Evolution, Ecology, Structure and Function Systems

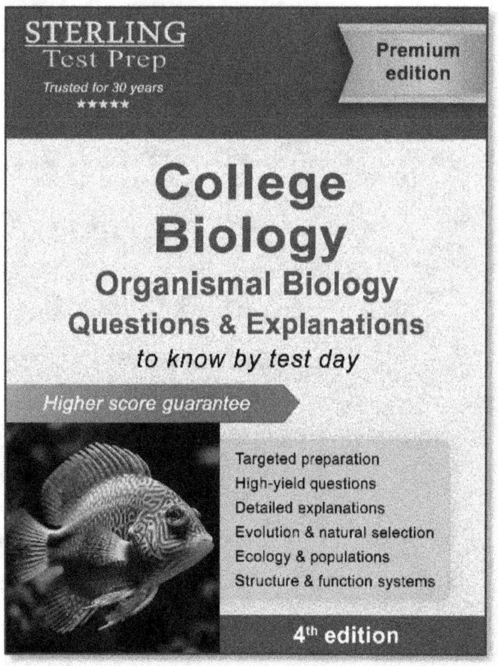

If you benefited from this book, please leave a review on Amazon so others can learn from your input. Reviews help us understand our customers' needs and experiences while keeping our commitment to quality.

CLEP study aids by Sterling Test Prep

Biology Review

Biology Practice Questions

Chemistry Review

Chemistry Practice Questions

Introductory Psychology

Introductory Sociology

American Government

History of the United States I

History of the United States II

Western Civilization I

Western Civilization II

Visit our Amazon store

www.ingramcontent.com/pod-product-compliance
Lightning Source LLC
LaVergne TN
LVHW080019210926
848481LV00011B/3057